Groundworks

Algebra Puzzles and Problems
Grade 7

T5-BAW-245

Carole Greenes
Carol Findell

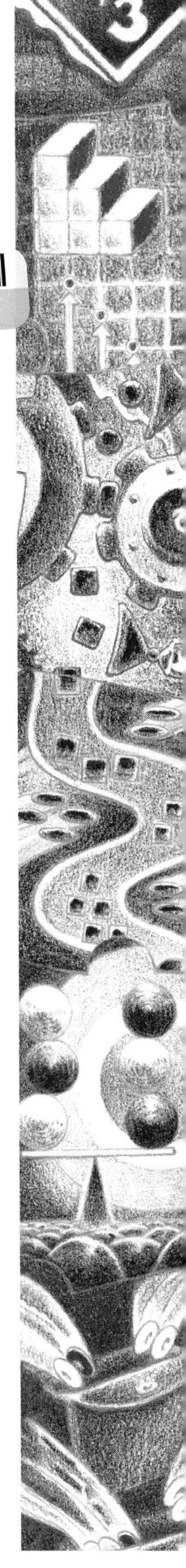

Creative Publications®

Design Director: Gregg McGreevy
Editor: Ann Roper
Art Director: Meg Saint-Loubert
Illustrator: Duane Bibby
Production Coordinator: Cindy Koller
Production Director: Ed Lazar
Designer: Carolyn Deacy Design
Production: University Graphics, Inc.
Manufacturing Coordinator: Michelle Berardinelli
Cover Illustrator: Susan Aiello

Unless otherwise stated on selected pages in this book, no part of this publication may be reproduced, or copied into or stored in a retrieval system, or transmitted, in any form or by any means, electronic, mechanical, photocopying, recording, or otherwise, without the prior written permission of the publisher.

Groundworks and Algebra Puzzles and Problems are trademarks of Creative Publications.

Creative Publications is a registered trademark of Creative Publications.

©1998 Creative Publications® Inc.
Two Prudential Plaza
Chicago, IL 60601

Printed in the United States of America

ISBN: 0-7622-0560-1

4 5 6 7 ML 05 04 03 02 01 00

Contents

Algebra: Puzzles and Problems

Why?

Algebra for Everyone is promoted by the National Council of Teachers of Mathematics (NCTM) in its Curriculum and Evaluation Standards (1989), by the College Board's Equity Project, and by authors of the SCANS Report (1991) on needs of the workplace. As a consequence, school districts around the country require all students to study algebra in high school. In some schools students take formal algebra courses as early as the seventh, or even sixth, grade. Although students are capable of succeeding in algebra, they often do not. Why aren't they successful?

From the authors' experience, even more able math students, jumping from an arithmetic-driven program directly into the study of algebra, often find the new content confusing and daunting. The main reason for this difficulty and for subsequent failure is lack of preparation. Although the NCTM has recommended that students gain experience with the big ideas of algebra during their elementary school years, current mathematics programs do not include such a preparatory program.

What?

Algebra: Puzzles and Problems contains problems for students in grades 4-7, to develop their understanding of six big ideas of algebra:

- representation
- proportional reasoning
- balance
- variable
- function
- inductive/deductive reasoning

The problems capitalize on students' experiences with arithmetic and arithmetic reasoning and help them make the connection between arithmetic and algebra. With such opportunities to make connections and to explore the big ideas of algebra before their formal study of the subject, students enhance their chances for success with algebra and with algebraic reasoning.

Big Ideas

Each problem in the book, though appearing under one big idea, generally requires the understanding of more than one algebraic idea.

Representation

Representation is the display of mathematical relationships in diagrams, drawings, graphs (bar, picto-, line, and scatter plot), symbols, tables, time lines, and text.

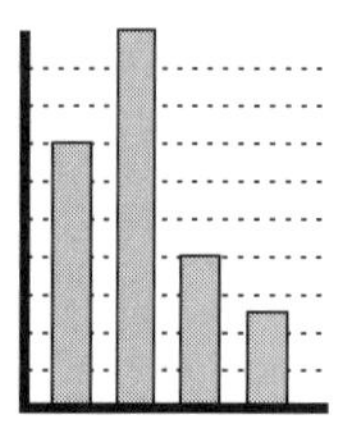
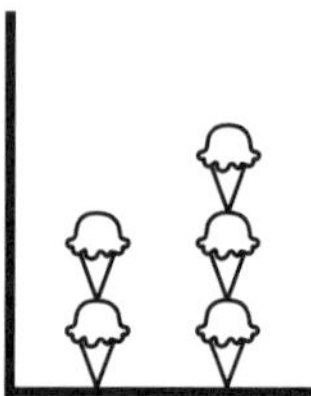
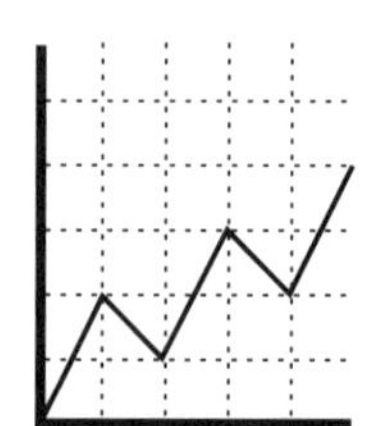
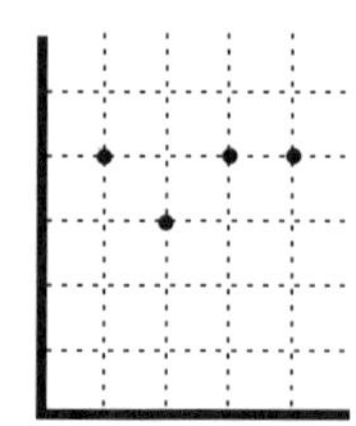

© Creative Publications

In this book, *What's the Point?* and *The Right Graph* are the families of problems that develop representation ability. Representation ability involves:

- interpretation of mathematical relationships shown in many forms
- matching of different representations of the same relationship; for example, descriptions in words matched with graphic displays
- creation of multiple representations of the same relationship
- recognition of how a change in one representation of a relationship effects a change in other representations of the same relationship

Proportional Reasoning

Determining how objects vary in relation to one another is the essence of proportional reasoning. Students use proportional reasoning when they interpret maps and scale drawings, determine dimensions of enlargements and reductions, compute unit costs, generate equivalent ratios, and identify percentages or parts of groups. To give students practice in this type of reasoning, this book includes two families of problems, *This and That* and *In Scale.*

Balance

Balance deals with the concept of equality among variable expressions. The problems in *Balance It* and *Place It Right* aid in the development of understanding of equality and ways to modify inequalities to achieve balance or equality. This is necessary preparation for work with solving equations and inequalities in algebra.

Variable

Fundamental to algebra is understanding the meaning of variable. Using variables is a way of generalizing relationships and representing quantities in formulas, functions, and equations, and of computing in expressions and equations. *Unknown Weights, Solve It, Puzzle Grid,* and *Tag Sale* help students to identify relationships among variables, to use variables in systems of equations, and to solve equations with variables, using the processes of substitution and replacement.

Function

Patterns and functions are important ideas in mathematics. A function is a relationship in which two sets are linked by a rule that pairs each element of the first set with exactly one element of the second set. There are three families of problems that deal with functions: *Make a Table, Beginning and End,* and *Functions and Graphs.* Students will also learn to describe in words or symbols rules for relating inputs and outputs and to construct inverse operation rules (to undo what other operations did).

Inductive Reasoning

Reasoning inductively requires the ability to examine particular cases, identify patterns and relationships among those cases, and extend the patterns and relationships. In the problem families *Keep Building* and *Where's the Seat?* students learn to recognize, extend, and generalize patterns, and to write rules in words or symbols to describe them.

Deductive Reasoning

Deductive reasoning involves the ability to infer a new relationship from information gleaned from two or more stated or displayed relationships. Most problems in this book require making inferences and drawing conclusions. Deductive reasoning is not one of the big ideas of this book; however, it is often one of the goals for a problem set.

What is in this Book?

Each Algebra: Puzzles and Problems book contains general teacher information, 30 blackline-master problem sets for students (three problems per set), questions, solutions, a Management Chart, and a Certificate of Excellence, both of which are also blackline masters.

Problem Sets

Each problem set consists of four pages. The first page presents the problem. On the facing page, there is teacher information, including goals listing specific mathematical reasoning processes or skills, questions to ask students, and solutions. The third page of each set presents two more problems dealing with the same big idea. Solutions are on the fourth page. Generally there is not room to show more than one solution method for each of these problems. If one method appears on a teacher page, another method may be shown for the second problem. Students can use either method for both problems. There are six problems in each family of problems.The mathematics required for all the problems is in line with the generally approved math curriculum for each grade level.

How to Use this Book

Have students work either individually or in pairs. Because many of the problem types will be new to your students, you may want to have the entire class or a large group of students work on the first problem in a set at the same time. You can use the questions that accompany the problem as the basis for a class discussion. As the students work on the problem, help them with difficulties they may encounter. To help them in their thinking, provide feedback such as, *How do you know? Does that seem reasonable? Explain your answer.* Once students have completed the first problem in each set, you can assign the next two problems for the students to do on their own.

Although the big ideas and the families of problems within them come in a certain order, your students need not complete them in this order. They might work the problems based on the mathematics content of the problems and their alignment with your curriculum, or according to student interests or needs.

One way to use this book is to have students work one set of problems per week for 30 weeks, in what is generally a 36-week school year. If you present and discuss the first problem in a set early in a week, the students could have the rest of the week to complete the other two problems, either in class or as homework.

Since the idea behind the problems is to give students experience with the problems, which focus to a great extent on reasoning and not on computation, it seems reasonable that your students use calculators if they wish to do so.

There is a Management Chart on page vii that you may duplicate for each student to keep in a portfolio. You may award The Certificate of Excellence (page 120) upon the successful completion of problem sets.

© Creative Publications

Management Chart

Name ______________________________ Class ______ Teacher ______________

BIG IDEA	PROBLEM SET				DATE
Representation	What's the Point?	A	B	C	
		D	E	F	
	The Right Graph	A	B	C	
		D	E	F	
Proportional Reasoning	This and That	A	B	C	
		D	E	F	
	In Scale	A	B	C	
		D	E	F	
Balance	Balance It	A	B	C	
		D	E	F	
	Place It Right	A	B	C	
		D	E	F	
Variable	Unknown Weights	A	B	C	
		D	E	F	
	Solve It	A	B	C	
		D	E	F	
	Puzzle Grid	A	B	C	
		D	E	F	
	Tag Sale	A	B	C	
		D	E	F	
Function	Make A Table	A	B	C	
		D	E	F	
	Beginning and End	A	B	C	
		D	E	F	
	Functions and Graphs	A	B	C	
		D	E	F	
Inductive Reasoning	Keep Building	A	B	C	
		D	E	F	
	Where's the Seat?	A	B	C	
		D	E	F	

© Creative Publications. Permission is given by the publisher to the purchasing teacher or parent to reproduce this page for classroom or home use only.

What's the Point? (A)

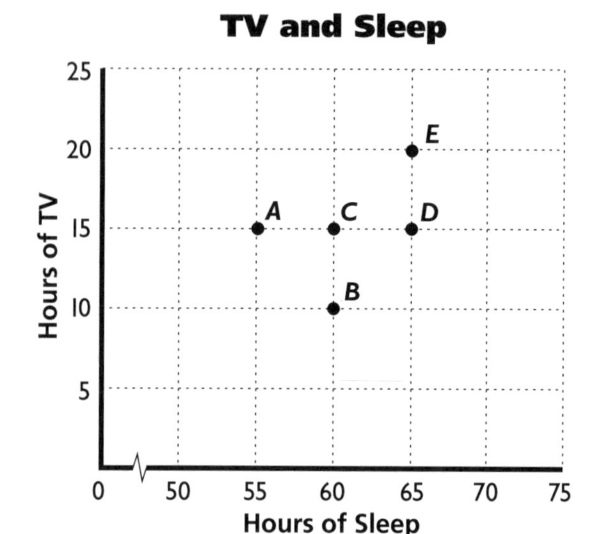

The scatter plot shows the weekly average number of hours of sleep and the weekly average number of hours of TV viewing of each of five friends.

Clues

a. Elyse watches fewer hours of TV than her friends.

b. Corey gets at least one more hour of sleep each night than Stacey.

c. Of all of them, Stacey gets the least sleep.

d. Leland watches more TV than Manuel does.

Tell which point represents which person.

1 A is ________

2 B is ________

3 C is ________

4 D is ________

5 E is ________

Name ....

Permission is given by the publisher to the purchasing teacher or parent to reproduce this page for classroom or home use only.

© Creative Publications

What's the Point? (A)

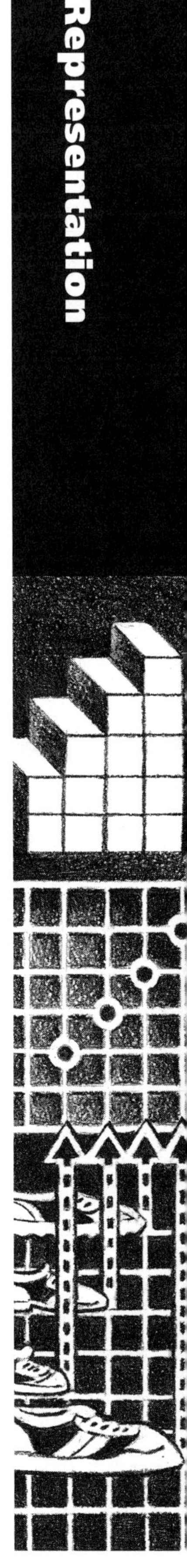

Goals

- Interpret a scatter plot.
- Match mathematical relationships presented in words with those shown in a graph.
- Make inferences.

Questions to Ask

- *Which point represents the person who watches the most TV?* (Point E)
- *How many people watch 15 hours of TV per week?* (3)
- *What is the fewest hours of sleep per week represented in this graph?* (55)
- *Which points represent people who get an average of 65 hours of sleep per week?* (D and E)
- *Which point represents someone who sleeps four hours for each hour of TV viewed?* (Point C)

Solutions

Clue a Since Elyse watches fewer hours of television than any of the others, Point B must represent her.

Clue c Stacey gets the least sleep; Point A is Stacey.

Clue b Corey gets at least one hour more of sleep each night than Stacey, so she gets at least 55 + 7, or 62, hours of sleep per week. So either Point D or E represents Corey.

Clue d Leland watches more TV than Manuel. They must be two of the points C, D, or E. Since Leland watches more TV than Manuel, and Points C and D represent people who watch the same number of hours of TV each week, Point E must represent Leland. Then Point D represents Corey and Point C, Manuel.

1 A is Stacey.

2 B is Elyse.

3 C is Manuel.

4 D is Corey.

5 E is Leland.

© Creative Publications

What's the Point? (B)

The scatter plot shows the average number of hours of sleep per week and the average number of hours of TV viewing per week of each of the five friends.

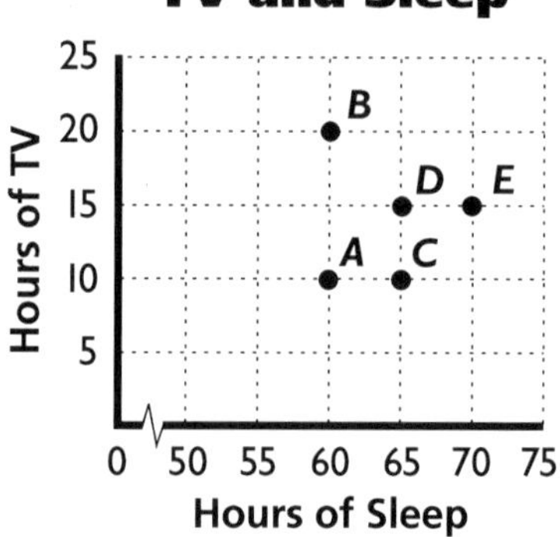

Clues

a. Juan and Angela watch TV the same number of hours per week.

b. Marita sleeps three hours for every hour of TV she watches.

c. Angela sleeps more hours than Yuri.

d. Cindy and Marita sleep the same number of hours.

Tell which point represents which person.

1 A is ______ **2** B is ______ **3** C is ______

4 D is ______ **5** E is ______

Name ....

Permission is given by the publisher to the purchasing teacher or parent to reproduce this page for classroom or home use only.

What's the Point? (C)

The scatter plot shows the average number of hours of sleep per week and the average number of hours of TV viewing.

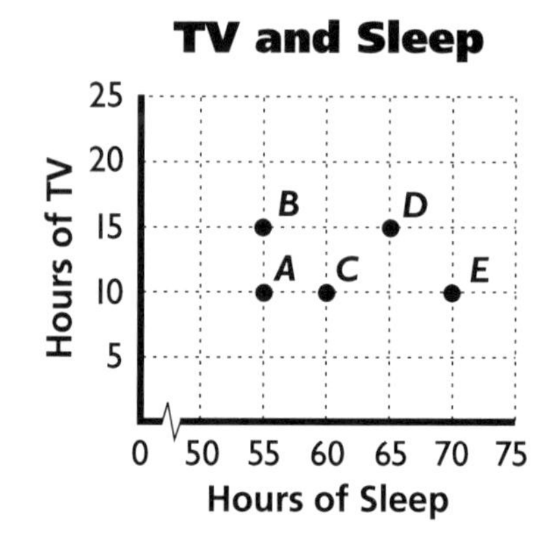

Clues

a. Leanne and Joe watch TV the same number of hours per week.

b. Aaron gets more sleep than the others.

c. Joe watches more TV than Aaron and LaToya.

d. Bjorn gets five more hours of sleep than Leanne.

Tell which point represents which person.

1 A is ______ **2** B is ______ **3** C is ______

4 D is ______ **5** E is ______

Name ....

Permission is given by the publisher to the purchasing teacher or parent to reproduce this page for classroom or home use only.

© Creative Publications

What's the Point? (B)

Solutions

Clue b Marita sleeps 3 hours for every hour of TV she watches. Point B represents Marita, since she watches TV 20 hours and sleeps 60 hours each week.

Clue d Since Cindy sleeps the same number of hours as Marita, Point A represents Cindy.

Clue a Points D and E represent Juan and Angela, who watch the same amount of TV.

Clue c Yuri must be Point C, the only point left. Since Angela sleeps more than Yuri, Angela is Point E. That leaves Point D to represent Juan.

1 A is Cindy.
2 B is Marita.
3 C is Yuri.
4 D is Juan.
5 E is Angela.

Representation

What's the Point? (C)

Solutions

Clue b Aaron sleeps more than the others, so Point E represents him.

Clue c Of the four points left, two represent 15 hours of TV per week, and two represent 10 hours. If Joe watches more TV than Aaron, Point B or D represents Joe.

Clue a Leanne and Joe watch the same amount of TV, so Leanne is Point B or D.

Clue d Either Point A or C represents Bjorn. Since he gets 5 more hours of sleep than Leanne, Point C represents him and Point B, Leanne. Point D represents Joe and Point A, LaToya.

1 A is LaToya.
2 B is Leanne.
3 C is Bjorn.
4 D is Joe.
5 E is Aaron.

© Creative Publications

What's the Point? (D)

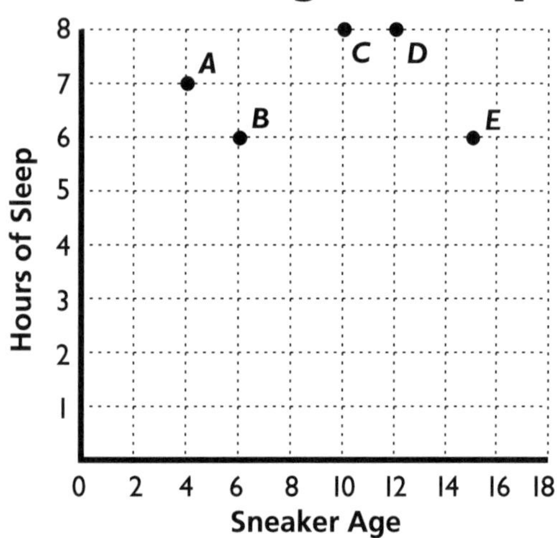

The scatter plot shows the age of sneakers in months and the average number of hours of sleep per night for five students.

Clues

a. The age of Brad's sneakers is a multiple of the ages of Diahanne's and Josh's sneakers.

b. The greatest common factor of the age of Linda's sneakers and the age of Amy's sneakers is five.

c. Linda sleeps about seven more hours per week than Diahanne.

Tell which point represents which person.

1 A is ______

2 B is ______

3 C is ______

4 D is ______

5 E is ______

Name ....

Permission is given by the publisher to the purchasing teacher or parent to reproduce this page for classroom or home use only.

© Creative Publications

What's the Point? (D)

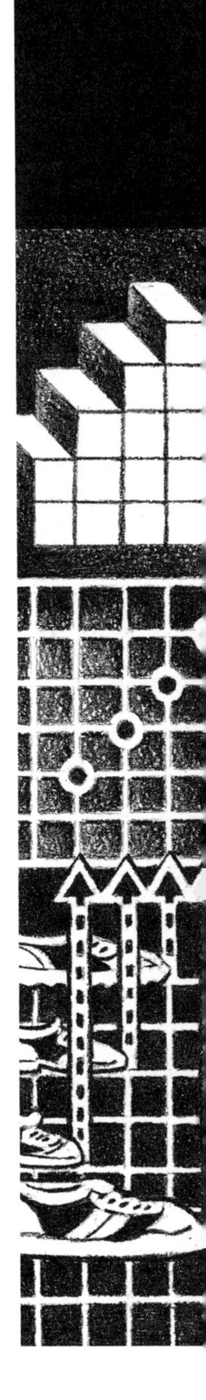

Goals

- Interpret a scatter plot.
- Match mathematical relationships presented in words with those shown in a graph.
- Make inferences.

Questions to Ask

- *What does each point represent?* (The average number of hours a student sleeps each night and the age in months of that student's sneakers)
- *Which point represents a student who sleeps an average of seven hours per night?* (Point A)
- *Which points represent students having sneakers that are more than ten months old?* (D and E)

Solutions

Clue a Since the age of Brad's sneakers is a multiple of the ages of 2 other students' sneakers, Brad's sneakers must be 12 months old. So Point D is Brad. Points A and B must represent Diahanne and Josh (sneakers are 4 and 6 months old).

Clue b Points C and E are left. Clue (b) names the other 2 students. Use the greatest common factor (5) as verification.

Clue c If Linda sleeps 7 more hours per week than Diahanne, then the difference in their nightly hours of sleep is 1 hour. Since either Point A or B is Diahanne, and C or E is Amy, find which point, C or E, is greater than A or B in number of hours of sleep. Since B and E both sleep 6 hours per night, and E sleeps less than A, C must represent Linda. Since Linda sleeps 8 hours per night, then A must represent Diahanne at 7 hours. Point B represents Josh; Point E is Amy.

1 A is Diahanne.

2 B is Josh.

3 C is Linda

4 D is Brad.

5 E is Amy.

What's the Point? (E)

The scatter plot shows the age of sneakers in months and the average number of hours of sleep per night for five students.

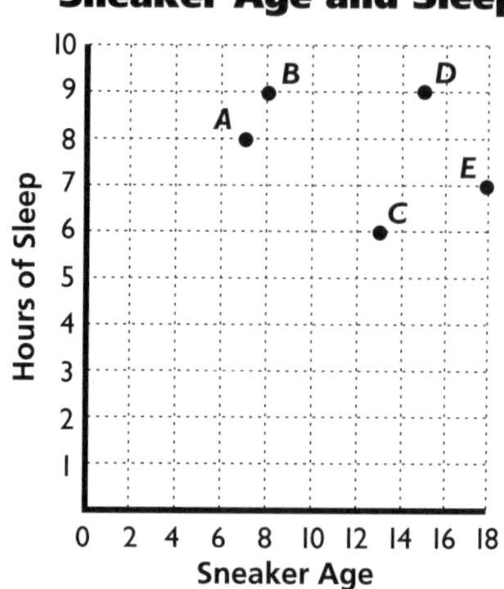

Clues

a. Anne sleeps one fewer hour each night than Jean.

b. The age of Ted's sneakers equals the total age of Sue's and of John's sneakers.

c. Jean's sneakers are two months more than twice the age of Sue's sneakers.

Tell which point represents which person.

1 A is ______ **2** B is ______ **3** C is ______

4 D is ______ **5** E is ______

Name ....

Permission is given by the publisher to the purchasing teacher or parent to reproduce this page for classroom or home use only.

What's the Point? (F)

The scatter plot shows the age of sneakers in months and the average number of hours of sleep per night for five students.

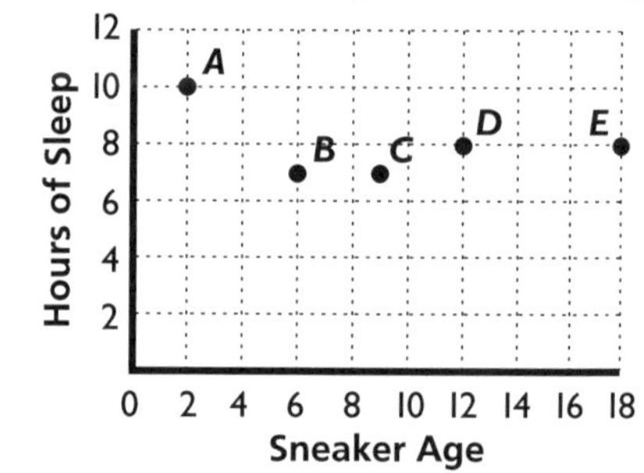

Clues

a. The age of Jerry's sneakers is the least common multiple of the ages of Leslie's and of Brenda's sneakers.

b. Chuck sleeps 7 hours more each week than Brenda and 14 hours fewer than Todd.

c. Brenda's sneakers are 4.5 times the age of Todd's sneakers.

Tell which point represents which person.

1 A is ______ **2** B is ______ **3** C is ______

4 D is ______ **5** E is ______

Name ....

Permission is given by the publisher to the purchasing teacher or parent to reproduce this page for classroom or home use only.

© Creative Publications

What's the Point? (E)

Solutions

Clue b Since the age of Ted's sneakers is the sum of the ages of Sue's and of John's sneakers, the only possibility is that Ted's sneakers are 15 months old. Point D represents him. 15 is the sum of 7 and 8. Points A and B represent Sue and John.

Clue a The only 2 points left are C and E. This clue names the other 2 students. Since Anne sleeps 1 fewer hour per night than Jean, Point C represents Anne; Point E represents Jean.

Clue c Jean's sneakers are 18 months old. Since they are 2 months more than twice the age of Sue's sneakers, and Sue is either A or B, compare 18 months to 7 months and to 8 months. Since $18 = 2 \times 8 + 2$, Point B represents Sue. That leaves Point A to represent John.

1 A is John.

2 B is Sue.

3 C is Anne.

4 D is Ted.

5 E is Jean.

Representation

What's the Point? (F)

Solutions

Clue c Brenda's sneakers are 4.5 times the age of Todd's sneakers. Only Points C and A are related in this way. So Point C represents Brenda, and A represents Todd; $4.5 \times 2 = 9$.

Clue b Chuck sleeps 1 hour more each night than Brenda. Since Point C represents Brenda who sleeps 7 hours per night, Chuck must sleep 8 hours per night; Point D or E must represent him. Chuck sleeps 2 fewer hours per night than Todd, who must sleep 10 hours per night.

Clue a The age of Jerry's sneakers is the LCM of the ages of Brenda's and of Leslie's sneakers. Brenda's sneakers are 9 months old, Jerry's sneakers must be 18 months old. Point E represents Jerry. That means Leslie's sneakers are 6 months old; Point B represents her. Since Point E represents Jerry, D must represent Chuck.

1 A is Todd.

2 B is Leslie.

3 C is Brenda.

4 D is Chuck.

5 E is Jerry.

Representation

© Creative Publications

The Right Graph (A)

Graph A

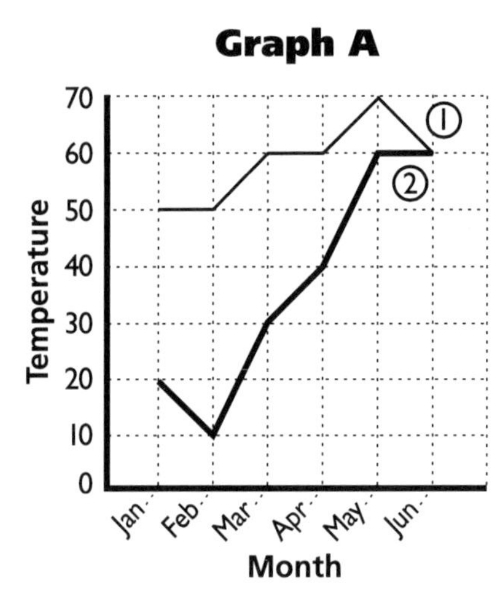

Graph B

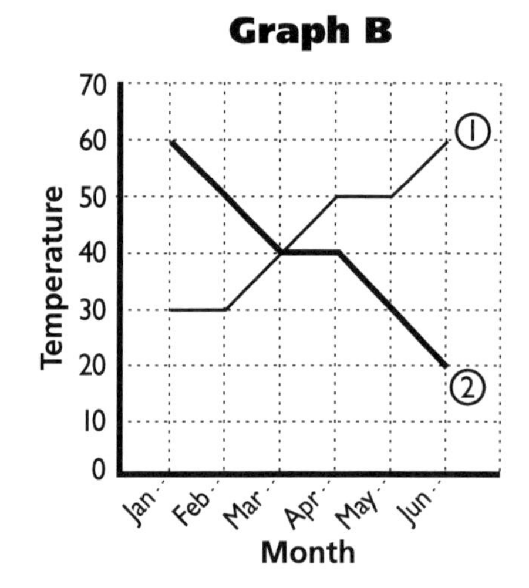

Each graph shows the average monthly temperatures in two cities for six months. It is never warmer in Salem than in Dover.

1 Which graph shows the average monthly temperatures in Salem and in Dover? Graph ______

2 Which line shows Salem's temperatures? Line ______

3 In which month is the temperature the same in both cities? ______ How does the graph show this?

4 In which city does the temperature vary more? ______

In January it is winter in Portsmouth, but summer in Sydney.

5 Which line represents Sydney's temperatures? Line ______

6 Which month is coldest in Portsmouth? ______

7 Which city has an average temperature of 40° for six months? ______

8 In which month is the difference in the temperatures between the two cities the greatest? ______

Name ....

Permission is given by the publisher to the purchasing teacher or parent to reproduce this page for classroom or home use only.

© Creative Publications

The Right Graph (A)

Goals

- Interpret double-line graphs.
- Compare graphs of linear relationships.
- Match line graphs to written descriptions of the same relationships.

Questions to Ask

- ***In Graph A which line represents the city with the warmer temperatures?*** (Line 1) ***How does the graph show this?*** (Line 1 is above Line 2 in every month except June, when the two lines meet.)
- ***In which month in Graph B did the two cities have the same average temperature?*** (March) ***How does the graph show this?*** (The two lines meet at this point.)

Solutions

1 Graph A - As the temperature increases, a line rises to show this. In Graph A, Line 1 is always above or at the same height as Line 2.

2 Line 2 - The line showing Dover's temperatures will always be above or at the same height as the line showing Salem's. Line 2 represents colder Salem.

3 June - The lines meet at this point.

4 Salem's temperature varies more. The distance between the least vertical and the greatest vertical height is greater for Line 2 than for Line 1.

5 Line 2 - Graph B shows the temperatures for Sydney and Portsmouth. The seasons for these two cities are different. For Line 1 the temperature increases from January to June. Line 2 temperatures decrease.

6 January and February are equally cold in Portsmouth.

7 Sydney - The mean temperature for Line 1: $(30 + 30 + 40 + 50 + 50 + 60) \div 6 = 43.3$. The mean temperature for Line 2: $(60 + 50 + 40 + 40 + 30 + 20) \div 6 = 40$.

8 June

Representation

© Creative Publications

The Right Graph (B)

Name

Each graph shows average monthly temperatures in two cities for six months. In August it is 20 degrees colder in Nome than in Hartford.

1 Which graph shows the average monthly temperatures in Hartford and in Nome? Graph ______

2 Which line represents Hartford's temperatures? Line ______

In July it is warmer in Boston than in San Francisco.

3 Which line represents Boston's temperatures? Line ______

4 Which city has the higher average temperature for the six months? ______

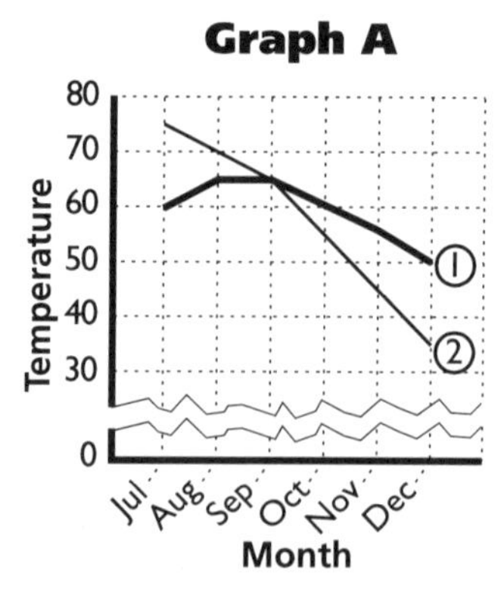

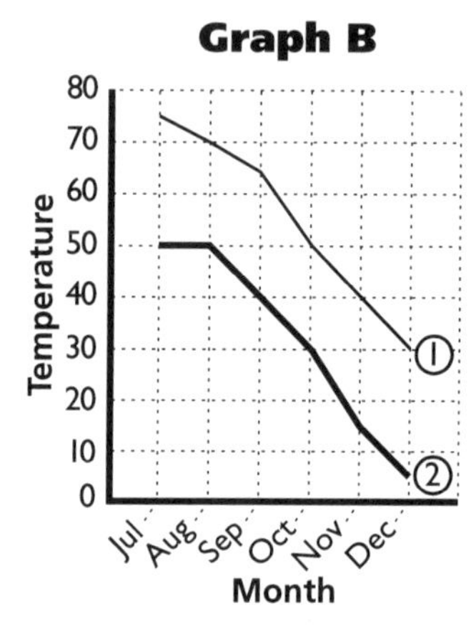

Permission is given by the publisher to the purchasing teacher or parent to reproduce this page for classroom or home use only.

The Right Graph (C)

Name

Each graph shows average monthly temperatures in two cities for six months. Honolulu's temperature varied by only five degrees for the six months.

1 Which graph represents temperatures in Honolulu and in Atlanta? Graph ______

2 Which line shows Atlanta's temperatures? Line ______

Anchorage was warmer than Juneau in only one of the six months.

3 Which line represents Juneau's temperatures? Line ______

4 In which two months did the two cities have the same average temperature? ______

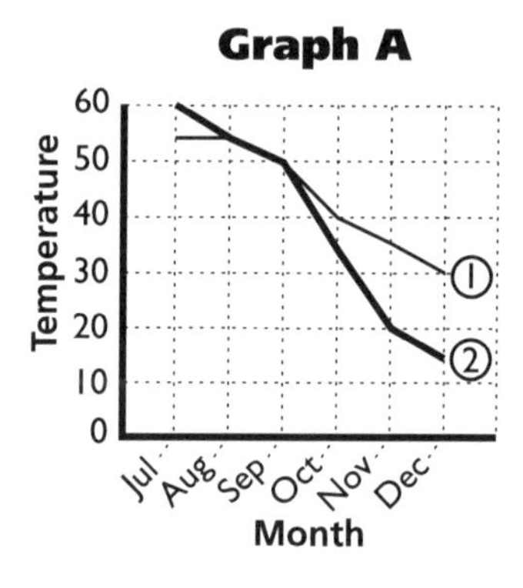

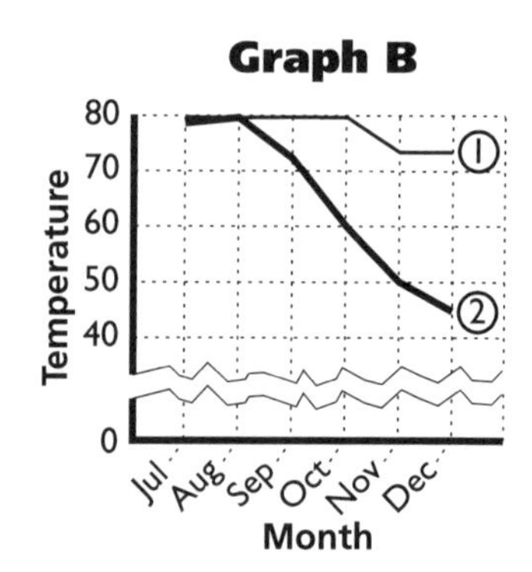

Permission is given by the publisher to the purchasing teacher or parent to reproduce this page for classroom or home use only.

© Creative Publications

The Right Graph (B)

Solutions

1 Graph B - In Graph A the vertical difference between the two lines at the point representing August is about 70 – 65, or 5°. In Graph B the vertical difference in August is 70 – 50, or 20°.

2 Line 1 - Since the temperature in August is colder in Nome, then Nome has the lower temperature and is represented by Line 2.

3 Line 2 - Graph A shows the temperatures in Boston and in San Francisco. In July the warmer temperature is shown by Line 2.

4 San Francisco - Mean temperature for Line 1: $(60 + 65 + 65 + 60 + 55 + 50) \div 6 = 59.2$

Mean temperature for Line 2: $(75 + 70 + 65 + 55 + 45 + 35) \div 6 = 57.5$

The mean temperature for the six months is higher in San Francisco, the city represented by Line 1.

The Right Graph (C)

Solutions

1 Graph B - Line 1 on Graph B is the only line that shows 5 degrees of variation over the six months.

2 Line 2 - Line 1 on this graph represents Honolulu's temperatures, so Line 2 represents Atlanta's.

3 Line 1 - Anchorage was warmer than Juneau during only one month. The Juneau-line is at the same height or above the Anchorage-line for all but one month.

4 August and September - The graphs are at the same height for these two months.

The Right Graph (D)

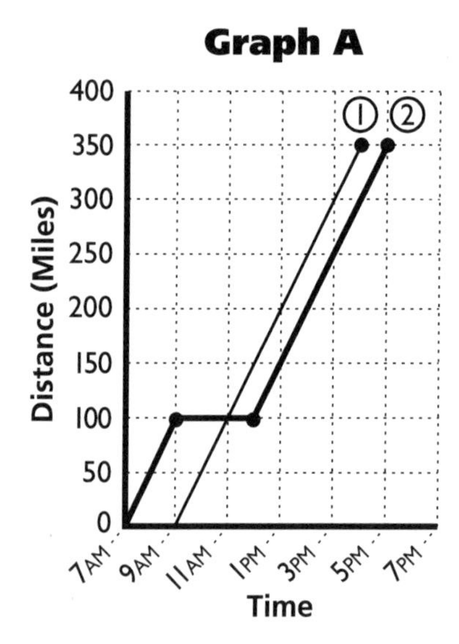

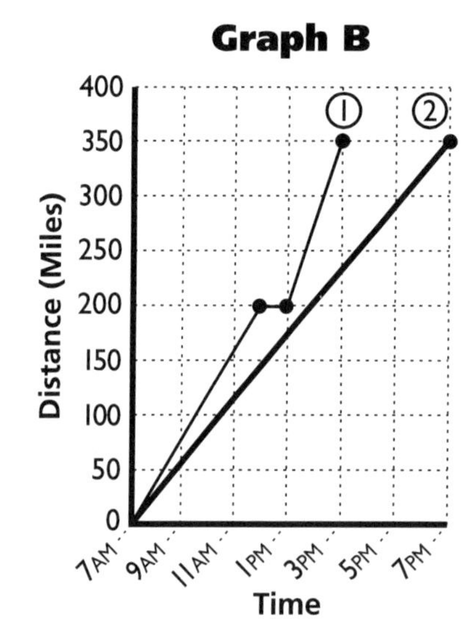

Each graph shows the trips of two families. The Li family left before the Rodriguez family. Both families took the same route.

1 Which graph shows their trips? Graph ______

2 Which line represents the Li's trip? Line ______

3 How many hours was the Rodriguez's trip? ______

4 Which family stopped to shop? ______
How does the graph show this?

The Clarks and the Jeffersons left on their trips at the same time. They took the same route.

5 The Clarks stopped for lunch. Which line represents the Clarks' trip? Line ______

6 At what time did the Jeffersons complete their trip? ______

7 Which family drove faster? ______
How can you tell from the graph?

8 Which family averaged more than 30 miles per hour? ______

Name ....

Permission is given by the publisher to the purchasing teacher or parent to reproduce this page for classroom or home use only.

The Right Graph (D)

Goals

- Interpret double-line graphs.
- Compare graphs of linear relationships.
- Match line graphs to text descriptions of the same relationships.

Questions to Ask

- *What does the horizontal axis show?* (Times of day at 2-hour intervals from 7 A.M. to 7 P.M.)
- *What does the vertical axis show?* (Distance in number of miles in 50-mile intervals)
- *On Graph B what time did the trip represented by Line 1 end?* (3 P.M.)
- *On Graph A what time did the trip represented by Line 2 begin?* (7 A.M.)

Solutions

1 Graph A shows two trips beginning at different times, one at 7 A.M. and the other at 9 A.M.

2 Line 2 - The Li family left before the Rodriguez group, at 7 A.M.

3 7 hours - Line 1 shows the Rodriguez's trip. They left at 9 A.M. and stopped at 4 P.M.

4 The Lis (Line 2) stopped. Part of the line, from 9 A.M. to noon, is horizontal and represents a stop, showing no change in distance.

5 Since the Clarks stopped for lunch, Line 1 on Graph B shows their trip. The horizontal segment between 12 and 1 P.M. represents the lunch stop.

6 Line 2 shows the Jeffersons finishing their trip at 7 P.M.

7 The Clarks (Line 1) drove faster. They went 350 miles in 8 hours (7 A.M. to 3 P.M.). It took the Jeffersons 12 hours to travel the same distance.

8 The Clarks - Line 1 shows 350 miles in 8 hours, about 44 miles per hour.

Representation

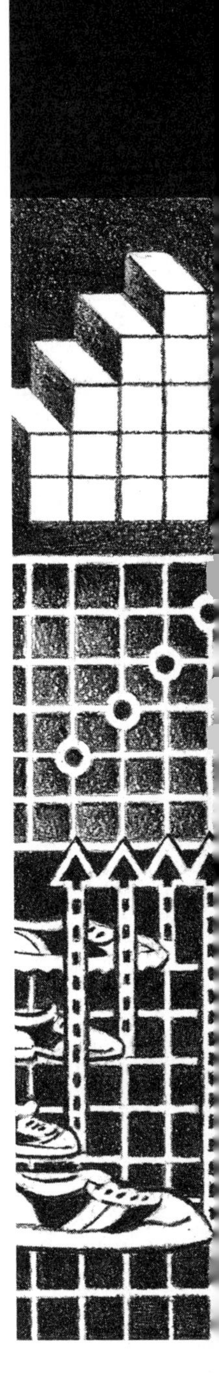

© Creative Publications

The Right Graph (E)

Each graph shows a bike race with two bikers. Jessie and Pedro started a bike race at the same time.

1 Which graph represents their race? Graph ______

2 Pedro traveled at a faster speed. Which line represents his race? Line ______

Carrie started the race after Marcie.

3 Which line represents Carrie's race? Line ______

4 Marcie started the race at 9 A.M. When did Carrie overtake Marcie? ______

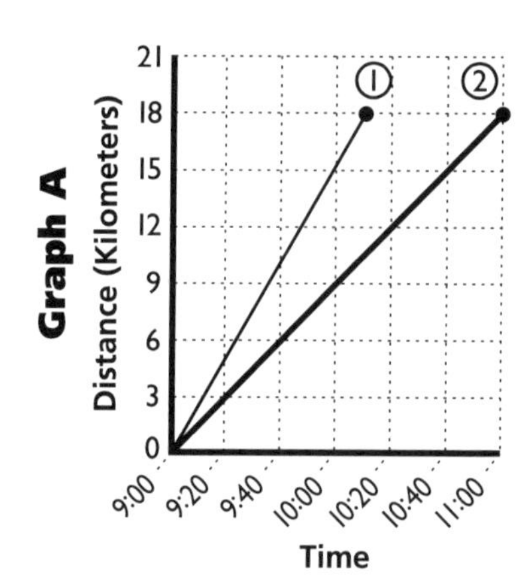

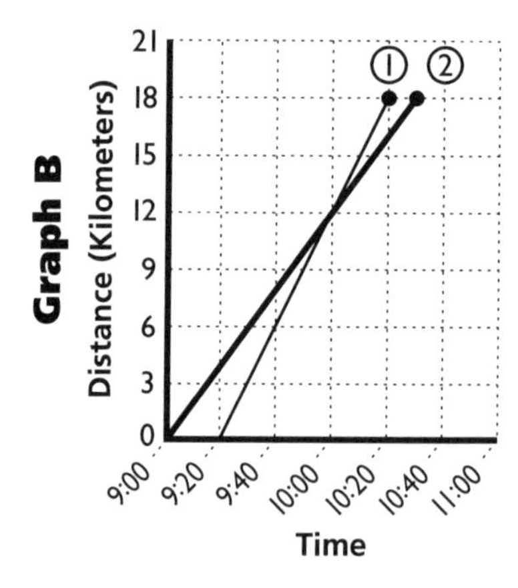

Name ....

Permission is given by the publisher to the purchasing teacher or parent to reproduce this page for classroom or home use only.

The Right Graph (F)

Each graph shows a foot race with two runners. In a foot race, Jed beat Alex to the finish line by 30 minutes.

1 Which graph represents Alex and Jed's race? Graph ______

2 At the same speed, how many minutes do you think it would take Jed to run 15 kilometers? ______

In the same race, Beth left 15 minutes before Elena.

3 Which line represents Elena's race? Line ______

4 How many minutes after Beth left did Elena overtake her? ______

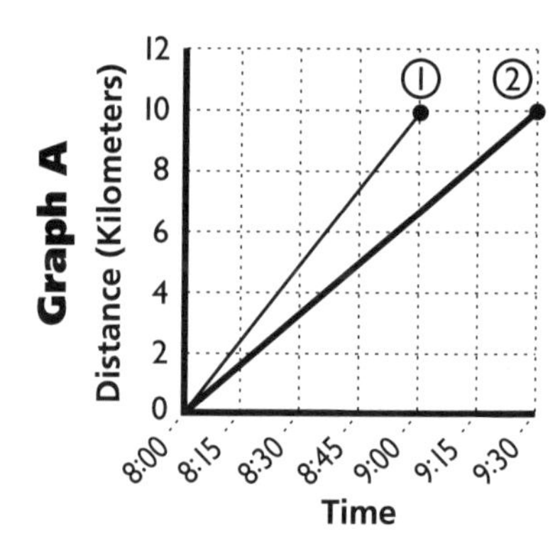

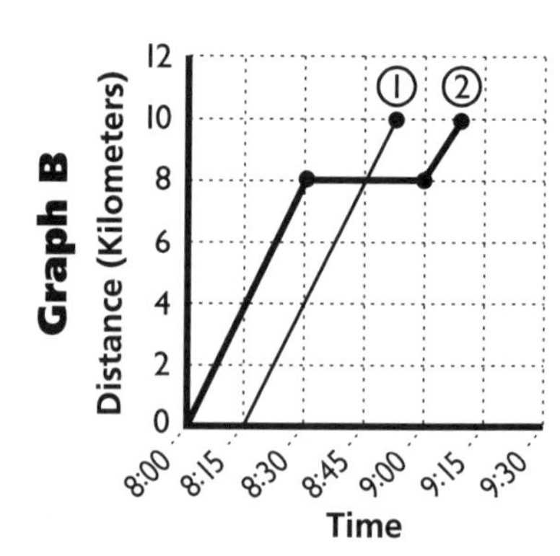

Name ....

Permission is given by the publisher to the purchasing teacher or parent to reproduce this page for classroom or home use only.

© Creative Publications

The Right Graph (E)

Solutions

1 Graph A - It shows two racers starting at the same time.

2 Line 1 - Pedro traveled faster than Jessie. Line 1 is steeper than Line 2; the steeper the line, the faster the speed. Line 1 shows a trip of 18 kilometers in 70 minutes, Line 2 shows a trip of 18 kilometers in 120 minutes. Since the distances are the same, compare the times; 70 is less than 120, so Line 1 represents the faster trip.

3 Line 1 - Since Carrie started after Marcie, Line 1 in Graph B represents her race.

4 At 10 A.M. the lines intersect. That means that 60 minutes after Marcie started, at a distance of 12 kilometers from the starting line, Carrie overtook Marcie.

The Right Graph (F)

Solutions

1 Graph A - It shows Jed finishing a half hour ahead of Alex.

2 90 minutes - Since Jed won and Line 1 represents his race, Jed ran 10 kilometers in 60 minutes. At that rate, Jed could run 15 kilometers in 90 minutes. Figure that 15 kilometers = 10 km + 5 km; 5 is half of 10. If it took Jed 60 minutes to run 10 km, he could run 5 km in 30 min. Ten km in 60 min + 5 km in 30 min is 15 km in 90 min.

3 Line 1 on Graph B shows Elena's race.

4 45 minutes - Overtake appears as the intersection of the lines. Elena overtook Beth 45 minutes after Beth started.

© Creative Publications

This and That (A)

Some studies show that in the United States, one out of seven people is left handed.

1 In an office of 28 people, how many people would you expect to be left handed? _______ Explain your answer.

2 If six people in an office are left handed, how many people would you expect to be right handed? _______ Explain your answer.

3 Fifty employees in a company are left handed. How many people do you think work in that company in all? _______

Name ....

Permission is given by the publisher to the purchasing teacher or parent to reproduce this page for classroom or home use only.

© Creative Publications

This and That (A)

Goals

- Use a proportion to solve a problem.
- Generate equivalent ratios.

Questions to Ask

- *If 1 out of 7 people in the U.S. is left handed, how many people out of 7 are right handed?* (6)
- *In a group of 14 people, how many people would you expect to be right handed?* (2) *How do you know?* (14 is two groups of 7. If 1 out of 7 is left handed, then 2 out of 14 are left handed.)
- *In a survey 18 people said they were right handed. How many people do you think were surveyed?* (Since 6 out of 7 people are right handed, then the total number of people surveyed was about 21.)

Solutions

1 Four people would be left handed. Set up and solve a proportion: $\frac{1}{7} = \frac{L}{28}$; $7L = 28$; $L = 4$.

Another way to solve the problem is to think: One-seventh of the group is left-handed, so $\frac{1}{7} \times 28 = 4$.

2 36 people - If 6 people in an office are left handed, then for every 1 left-handed person there are 6 right-handed people. The number of right-handed people should be 6×6, or 36.

Another possible solution method:

Set up and solve a proportion. Use the proportion to determine the total number of people surveyed if you know that 6 of that total are left handed: $\frac{6}{T}$ (total) $= \frac{1}{7}$; $42 = T$. Reason that, since there are 42 people and 6 are left handed, $42 - 6$, or 36, are right handed.

3 350 people - If 50 employees are left handed and the ratio of left handed to total number of employees is $\frac{1}{7}$, then $\frac{1}{7} = \frac{50}{T}$; $T = 350$.

© Creative Publications

This and That (B)

In the village of Mill Run, 1 out of every 3 people lives in an apartment. The population of Mill Run is 360.

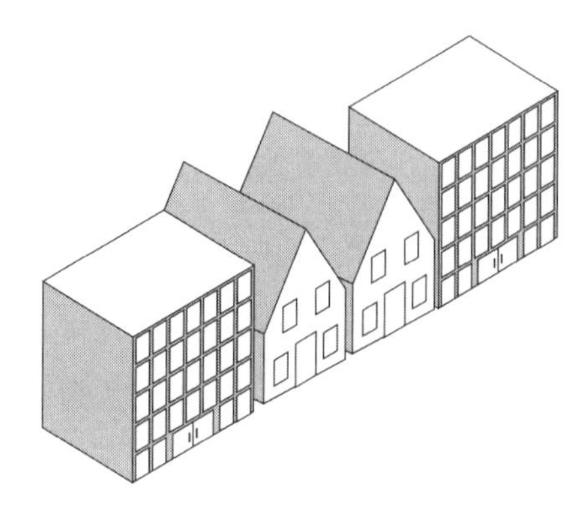

1 How many people in Mill Run live in apartments? _______ How did you find the answer?

2 How many people in Mill Run do not live in apartments? _______ Tell how you know.

3 Of the people who do not live in apartments, 1 out of every 2 people owns his own home. How many people own their own homes? _______ Record the steps you followed to get the answer.

Name ....

Permission is given by the publisher to the purchasing teacher or parent to reproduce this page for classroom or home use only.

This and That (C)

The population of Bloomfield is 4000. In the town of Bloomfield 2 out of 5 people ride buses to work.

1 How many people in Bloomfield ride buses to work? _______ How do you know?

2 How many people in Bloomfield do not ride buses to work? _______ How do you know?

3 Of the people who do not ride buses, 3 out of 4 ride in cars to work. How many people ride in cars? _______ Record the steps you followed to get the answer.

Name ....

Permission is given by the publisher to the purchasing teacher or parent to reproduce this page for classroom or home use only.

© Creative Publications

This and That (B)

Solutions

1 120 people - If 1 out of 3 people lives in an apartment, then $\frac{1}{3} \times 360$, or 120, people live in apartments. Another possible method for finding the number of people who live in apartments is to set up and solve a proportion: $\frac{1}{3} = \frac{P}{360}$; $3P = 360$; $P = 120$.

2 240 people - If there are 360 people in Mill Run and 120 live in apartments, then 240 don't. Or, write and solve a proportion: $\frac{2}{3} = \frac{N}{360}$; $720 = 3N$; $N = 240$.

3 120 people - To find the number of people who own their own homes, set up a proportion. Let H equal number of homeowners and T equal total number of people: $\frac{1}{2} = \frac{H}{240}$; $2H = 240$; $H = 120$. An easier method is to think: Half of the people who don't live in apartments own their own homes; half of 240 is 120.

Proportional Reasoning

This and That (C)

Solutions

1 1600 people - Since 2 out of 5 people ride buses to work, then $\frac{2}{5} = \frac{P}{4000}$; $5P = 8000$; $P = 1600$.

2 2400 people - Two possible solution methods:

If 1600 people ride buses, then 4000 – 1600, or 2400, don't ride buses.

Set up a proportion. Let D equal the number of people who don't ride buses. If 2 out of 5 people ride buses, then 3 out of 5 don't. $\frac{3}{5} = \frac{D}{4000}$; $5D = 12{,}000$; $D = 2400$.

3 1800 people - The number of people who don't ride buses is 2400. If 3 out of 4 ride in cars, then $\frac{3}{4} = \frac{P}{2400}$; $7200 = 4P$; $P = 1800$.

Proportional Reasoning

© Creative Publications

This and That (D)

In Chelsea Middle School, the ratio of the number of eighth-grade students taking French to the number of students not taking French is 1:4. The ratio is the same for both boys and girls.

1 If there are 240 eighth-grade students in Chelsea Middle School, how many take French? _______

2 Half the students in the eighth grade are girls. How many eighth-grade girls do not take French? _______ How do you know?

3 Five students in Sean's eighth-grade homeroom take French. How many students do you think are in Sean's homeroom? _______

Name ....

Permission is given by the publisher to the purchasing teacher or parent to reproduce this page for classroom or home use only.

 © Creative Publications

This and That (D)

Goals

- Use a proportion to solve a problem.
- Generate equivalent ratios.

Questions to Ask

- ***The numbers in the ratio 1:4 are both parts of a whole. What does the whole represent?*** (Total number of eighth-grade students)
- ***If the part-part ratio is 1:4, what is the corresponding part-whole ratio?*** (1:5)
- ***If 100 eighth-grade students take French, how many students do not take French?*** (400) ***How do you know?*** (The ratio is 1:4; 100:400 is an equivalent ratio.)
- ***In a group of 100 students, how many do you think take French?*** (20) ***How did you figure it out?*** (Since the part-part ratio is 1:4, then 1 of every 5 students takes French. One fifth of the students take French, and $\frac{1}{5}$ of 100 is 20.)

Solutions

1 48 students take French.

One possible solution method:

The ratio of students who take French to those who do not is 1:4. For every student who takes French, there are 4 who do not. For every 5 students, 1 takes French and 4 do not. Thus $\frac{1}{5}$ of the students take French, and $\frac{4}{5}$ of the students do not; $\frac{1}{5}$ of 240 is 48.

2 96 girls - Of the 240 students in grade 8, 120, or half, are girls. If $\frac{1}{5}$ of the girls take French, then the remaining $\frac{4}{5}$ do not. Use a proportion to find the number not taking French. Let X represent that number of girls not taking French: $\frac{4}{5} = \frac{X}{120}$; $5X = 480$; $X = 96$.

3 25 students - If 5 students take French, for each student taking French, 4 students do not. So 4×5, or 20, students do not take French. The total number of students in Sean's homeroom would be 25.

Proportional Reasoning

© Creative Publications

This and That (E)

In Chelsea Middle School, the ratio of the number of students in every grade who speak Spanish to the number of students who do not speak Spanish is 2:3.

1 In the seventh grade, 100 students speak Spanish. How many seventh-grade students don't speak Spanish? _______ How do you know?

2 If there are 1000 students in Chelsea Middle School, how many speak Spanish? _______ Tell how you know.

3 In Juan's class 18 students don't speak Spanish. How many students do you think are in the class? _______

Name ....

Permission is given by the publisher to the purchasing teacher or parent to reproduce this page for classroom or home use only.

This and That (F)

In Chelsea High School, the ratio of the number of students who take part in sports to the number of students who do not is 3:5.

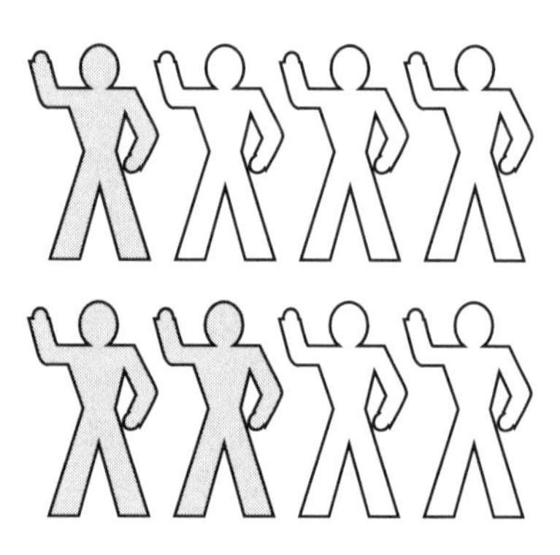

1 In the sophomore class, 250 students do not take part in sports. How many sophomores would you expect to take part in sports? _______ How do you know?

2 There are 408 students in Chelsea High School who take part in sports. How many students are there in the school? _______ How do you know?

3 The ratio of male to female students who take part in sports in Chelsea High School is 2:1. How many female sports players are in Chelsea High? _______ Explain your answer.

Name ....

Permission is given by the publisher to the purchasing teacher or parent to reproduce this page for classroom or home use only.

© Creative Publications

This and That (E)

Solutions

1 150 students - The ratio of Spanish-speaking to not-speaking is 2:3. Set up a proportion. Let N represent the number of students who don't speak Spanish: $\frac{2}{3} = \frac{100}{N}$; $2N = 300$; $N = 150$.

2 400 students - Since the part-part ratio is 2:3, the part-whole ratio is 2:5. To find the number of students who speak Spanish, set up a proportion. Let S represent the number of students who speak Spanish: $\frac{2}{5} = \frac{S}{1000}$; $5S = 2000$; $S = 400$.

3 30 students - Use a proportion. Let S represent the number of students in Juan's class: $\frac{3}{5} = \frac{18}{S}$; $3S = 90$; $S = 30$.

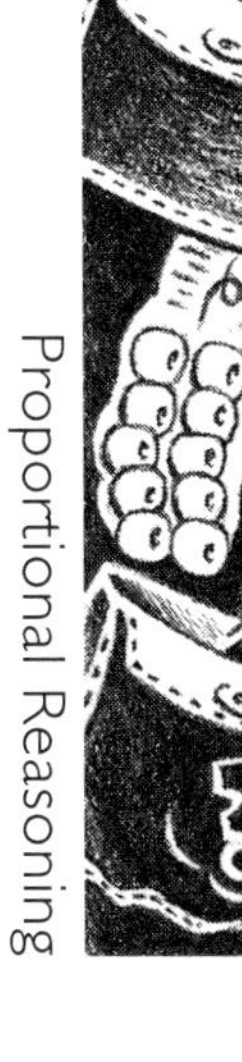

Proportional Reasoning

This and That (F)

Solutions

1 150 sophomores - The part-part ratio is 3:5. Use a proportion. Let S represent the number of students who take part in sports: $\frac{3}{5} = \frac{S}{250}$; $5S = 750$; $S = 150$.

2 1088 students - If the part-part ratio is 3:5, the corresponding part-whole ratio is 3:8. Use a proportion. Let S represent the number of students in the school: $\frac{3}{8} = \frac{408}{S}$; $3S = 3264$; $S = 1088$.

3 136 female sports players - There are 408 students in the school who take part in sports. Of these, the ratio of male to female is 2:1. That means the ratio of female to male athletes is 1:2. The corresponding part-whole ratio is 1:3. Use a proportion. Let F represent the number of females: $\frac{1}{3} = \frac{F}{408}$; $3F = 408$; $F = 136$.

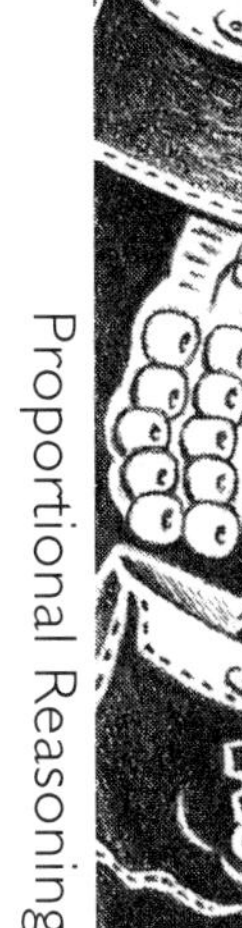

Proportional Reasoning

In Scale (A)

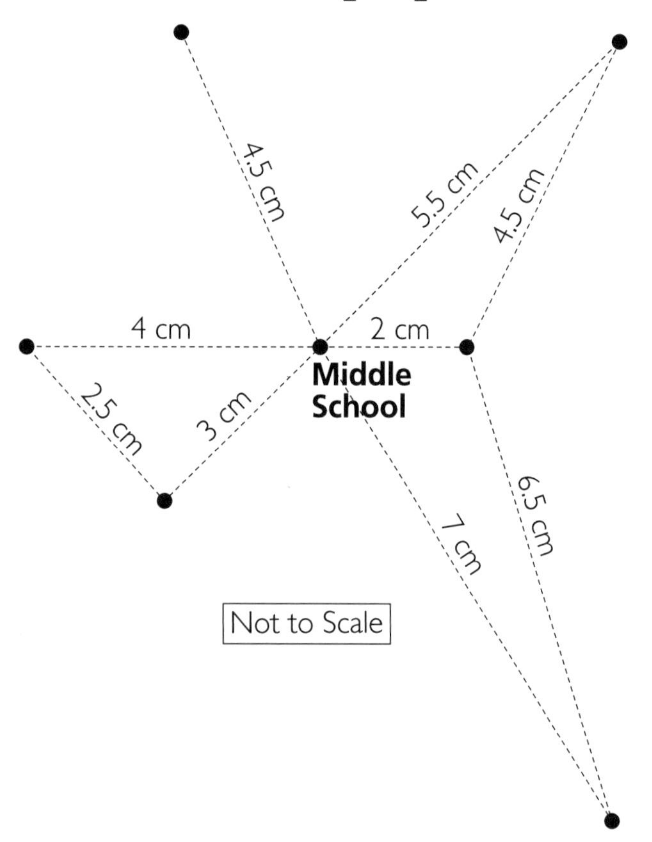

On the map 2 centimeters stand for 5 kilometers. Use the clues. Write the name of each building next to the dot on the map.

Clues

a. The distance from the middle school to the library by way of the high school is the same as the shortest distance from the high school to the zoo.

b. The shortest distance from the middle school to City Hall is 10 kilometers.

c. The fire station is 11.25 kilometers from the middle school.

d. The police station is 7.5 kilometers from the middle school.

Name

Permission is given by the publisher to the purchasing teacher or parent to reproduce this page for classroom or home use only.

© Creative Publications

In Scale (A)

Goals

- Match written relationships with those shown in maps.
- Interpret map scales.
- Write proportions to find real distances when given scale distances on a map.

Questions to Ask

- *What distance does 2 centimeters represent on the map?* (5 kilometers)
- *What does a map distance of 6 centimeters represent?* (15 kilometers) *How could you figure it out?* (You might write a proportion: $\frac{2 \text{ cm}}{5 \text{ km}} = \frac{6 \text{ cm}}{N \text{ km}}$; $2N = 30$; $N = 15$)
- *How many centimeters represent 17.5 kilometers?* (7 cm)

Solutions

One possible solution method:

Convert distances (in centimeters) between buildings to actual distances in kilometers, using proportions. Solve the proportions.

Clue d The police station is 3 centimeters from the middle school.

Clue b City Hall is 4 centimeters from the middle school.

Clue c The fire station is 4.5 centimeters from the middle school.

Clue a The 2 centimeters to the high school plus the 4.5 centimeters to the library is the same as the 6.5 centimeters from the high school to the zoo.

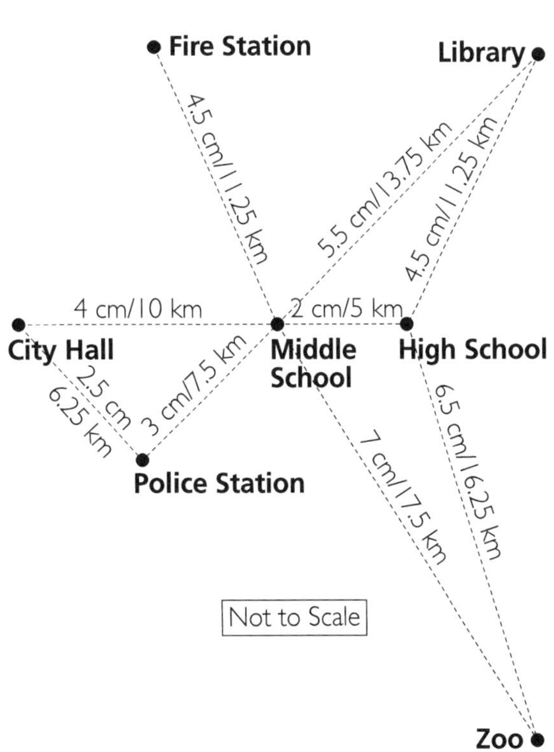

In Scale (B)

On the map 3 centimeters stand for 20 meters. Use the clues. Write the name of each ride next to the dot on the map.

Clues

a The distance from the Ferris wheel to the loop-the-loop is 40 meters.

b. The merry-go-round is 30 meters from the entrance.

c. A round trip from the Ferris wheel to the roller coaster is 60 meters.

d. The water slide is 50 meters from the roller coaster.

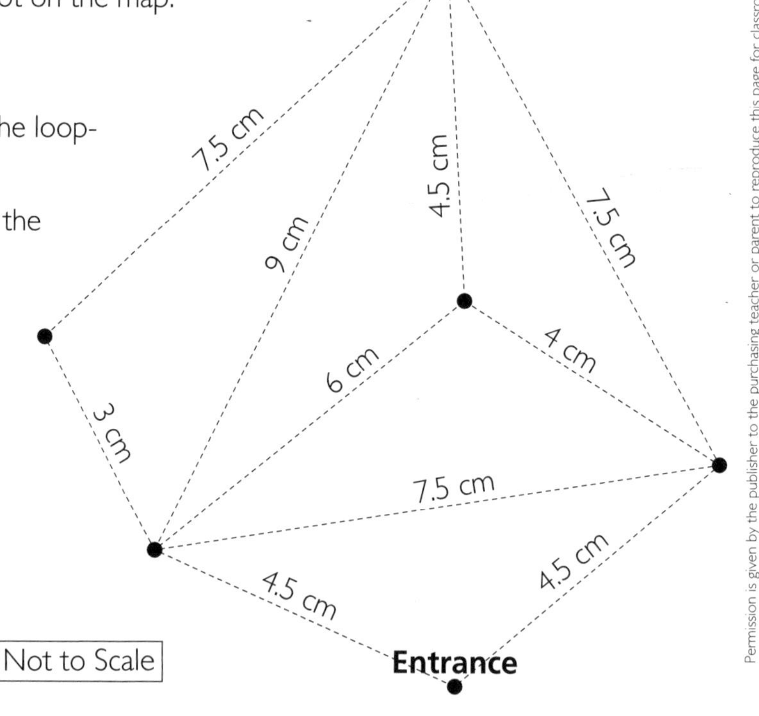

Name

In Scale (C)

On the map 5 centimeters stand for 8 meters. Use the clues. Write the name of each object next to the dot on the map.

Clues

a. From the house, to the swings, to the lilac bush, is 8 meters.

b. The trip from the tree house to the house, then to the swings, and back to the house is 16 meters.

c. From Big Rock to the swings is 8 meters. If you go by way of the lilac bush, the trip is almost a meter longer.

d. The distance from the house to the brook is the same as from the tree house to the brook.

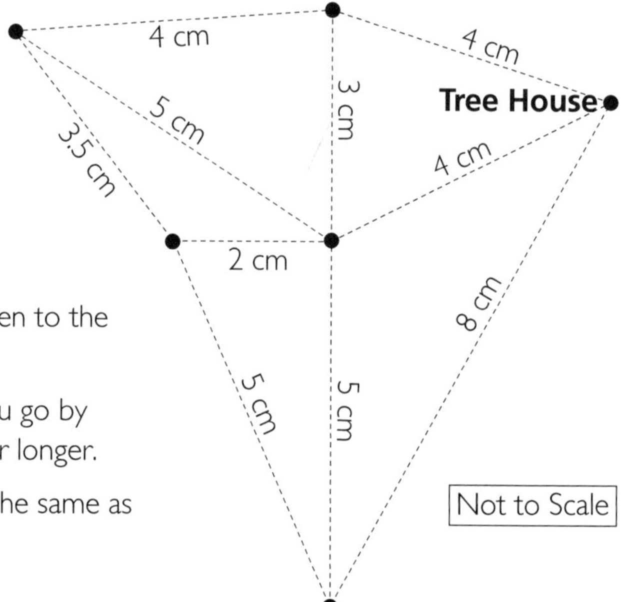

Name

Permission is given by the publisher to the purchasing teacher or parent to reproduce this page for classroom or home use only.

© Creative Publications

In Scale (B)

Solutions

Convert each map distance between rides to actual distances in meters, using proportions. Solve the proportions.

Clue b The merry-go-round is one of the 2 points 4.5 centimeters from the entrance.

Clue c Only the Ferris wheel and the roller coaster are 4.5 centimeters (30 meters) apart.

Clue a The Ferris wheel and the loop-the-loop must be 6 centimeters apart. This clue locates the Ferris wheel and the loop-the-loop. With the information from Clue (b), locate the roller coaster and the merry-go-round.

Clue d The water slide is 7.5 centimeters from the roller coaster.

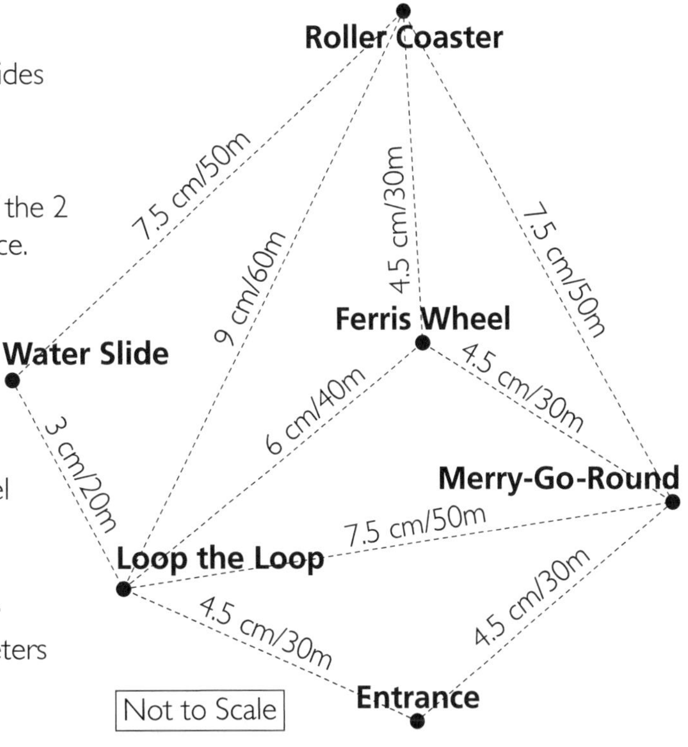

In Scale (C)

Solutions

Convert each map distance between places to actual distances in meters, using proportions. Solve the proportions.

Clue a Locate 2 distances that total 8 meters. These are 4.8 meters and 3.2 meters. The swings are in the center, and the house and lilac bush are at the ends.

Clue b The distance from the tree house, to the house, to the swings, and back is 16 meters. (6.4 + 4.8 + 4.8 = 16)

Clue c There are 2 points located 8 meters from the swings, but only 1 of them makes the trip via the lilac bush 1 meter longer.

Clue d Identify the remaining point as the brook. Check to see if the brook-to-tree-house distance is the same as the house-to-brook distance.

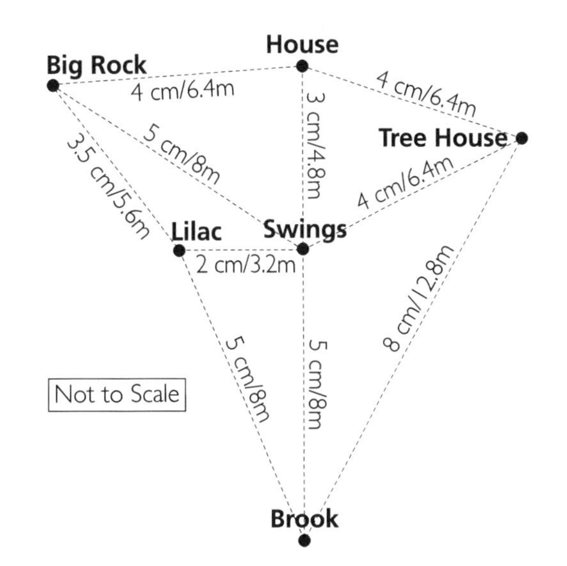

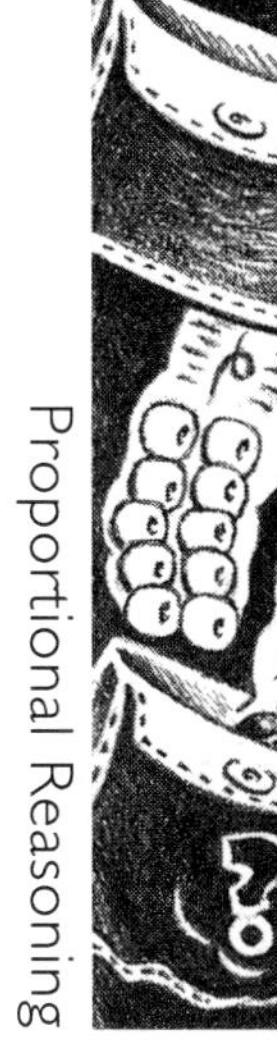

In Scale (D)

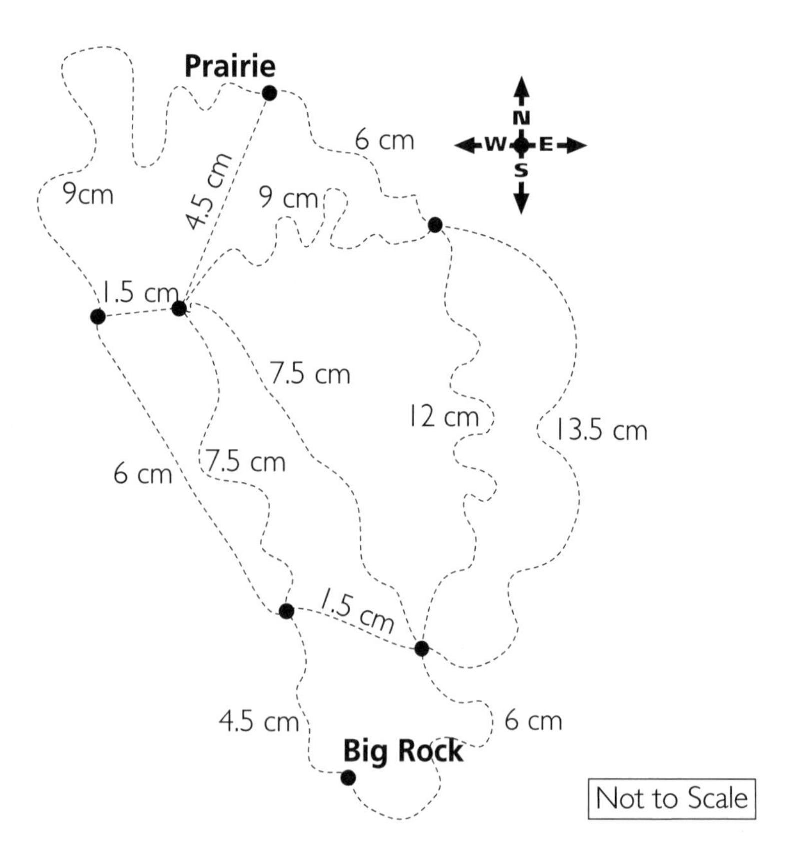

On the map 3 centimeters stand for 70 kilometers. Use the clues. Write the name of each city next to the dot on the map.

Clues

a. The distance from Big Rock to Fox Run is 35 kilometers greater than the distance from Big Rock to Trenton. The sum of the distances from Big Rock to Fox Run and from Big Rock to Trenton is 245 kilometers.

b. The shortest distance from Prairie to Trenton is by way of Canton. The distance from Prairie to Canton to Trenton is 268.3 kilometers.

c. Two round trips from Akron to Canton cover 140 kilometers, the same distance as one trip from Akron to Trenton.

d. The shortest distance from Edward to Trenton is 315 kilometers.

Name ....

Permission is given by the publisher to the purchasing teacher or parent to reproduce this page for classroom or home use only.

© Creative Publications

In Scale (D)

Goals

- Match relationships presented in words with those shown on maps.
- Interpret map scales.
- Write proportions to find real distances when given scale distances on a map.

Questions to Ask

- *On the map what distance does 3 centimeters represent?* (70 kilometers)
- *What does a map distance of 9 centimeters represent?* (210 kilometers) *How do you know?* (Use a proportion or just think: If 3 cm = 70 km, then 3 × 3, or 9, cm = 3 × 70 km.)
- *How many centimeters represent 140 kilometers on the map?* (6)

Solutions

One possible solution method:

Convert each map distance (in centimeters) between cities to actual distances (kilometers) by using proportions. For example, the map distance between Akron and Prairie is 9 cm. Find the number of kilometers by solving the proportion: $\frac{3 \text{ cm}}{70 \text{ km}} = \frac{9 \text{ cm}}{N \text{ km}}$; $3N = 630$; $N = 210$. The distance from Akron to Prairie is 210 km.

Compute all the distances in kilometers, then use the clues to identify the cities.

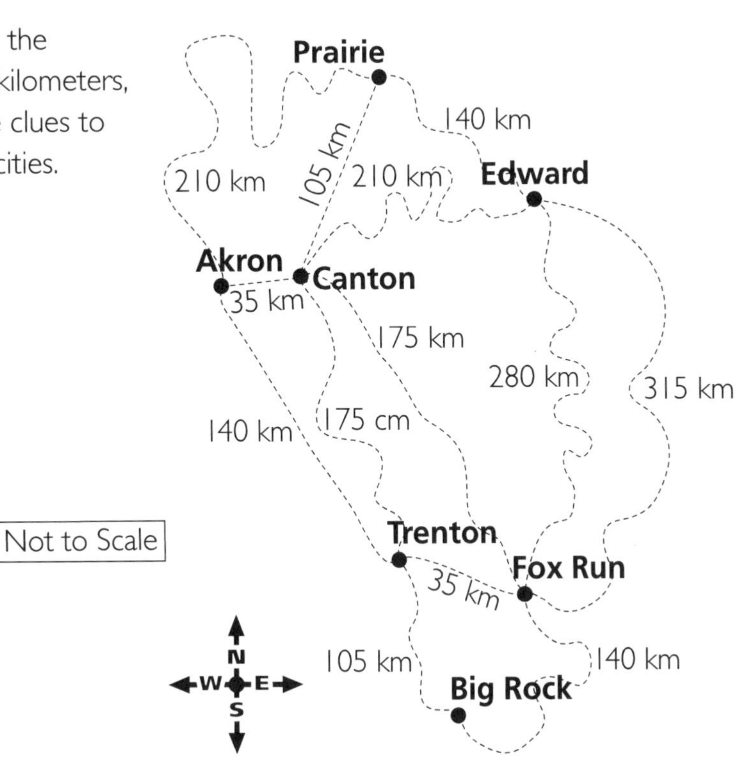

Proportional Reasoning

© Creative Publications

In Scale (E)

Name

On the map 5 centimeters stand for 40 kilometers.
Use the clues. Write the name of each city next to the dot on the map.

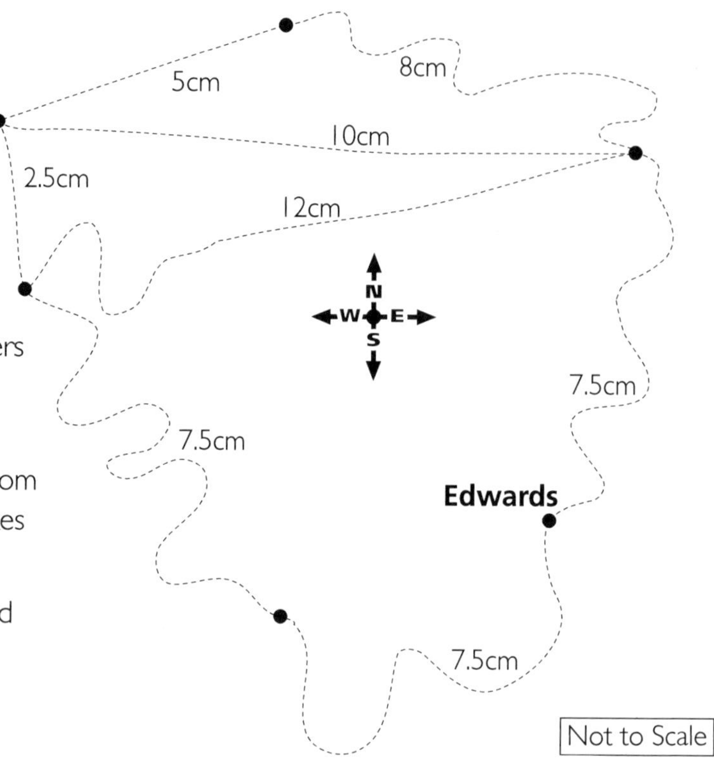

Clues

a. The trip from Delray to Leadville by way of Palm Village is 24 kilometers longer than the distance from Delray directly east to Leadville.

b. At a speed of 90kph, the trip west from Edwards to Warren to Sand City takes 80 minutes.

c. The round trip from Leadville to Sand City is 192 kilometers.

Permission is given by the publisher to the purchasing teacher or parent to reproduce this page for classroom or home use only.

In Scale (F)

Name

On the map 2 centimeters stand for 3 kilometers.
Use the clues. Write the name of each city next to the dot on the map.

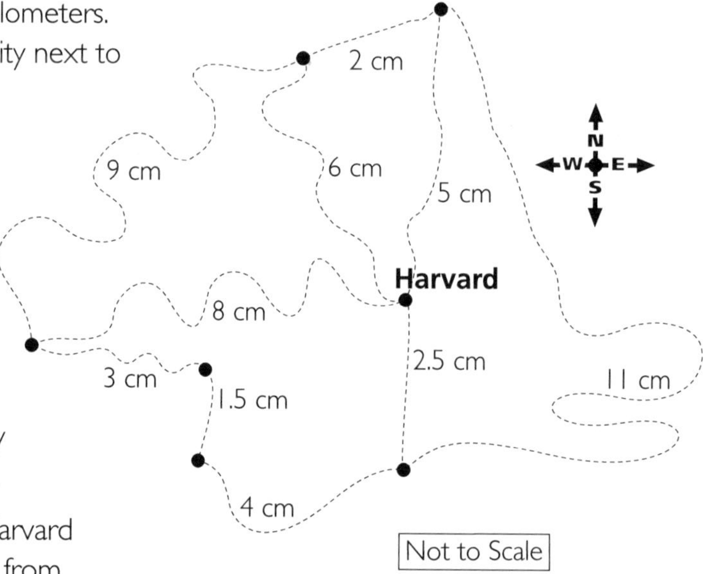

Clues

a. The shortest trip from Bayard to Harvard is 12 kilometers. On the trip you pass through Cedar Grove and then Branton.

b. At a speed of 30kph, the round trip from Waban to Branton the long way and back takes 1 hour and 6 minutes.

c. The trip from Waban to Saugus to Harvard is 4.5 kilometers shorter than the trip from Waban to Harvard to Saugus.

d. Hancock is 12 kilometers west of Harvard.

Permission is given by the publisher to the purchasing teacher or parent to reproduce this page for classroom or home use only.

© Creative Publications

In Scale (E)

Solutions

Use the ratio 5 cm/40 km and a proportion to compute the distance between cities in kilometers. $\frac{5}{40} = \frac{1}{8}$ or 1 cm represents 8 km.

Use a ratio to compute distances in kilometers. For example, to find the number of kilometers between Sand City and Warren, set up and solve a proportion: $\frac{1 \text{ cm}}{8 \text{ km}} = \frac{7.5 \text{ cm}}{N \text{ km}}$; $N = 60$ km.

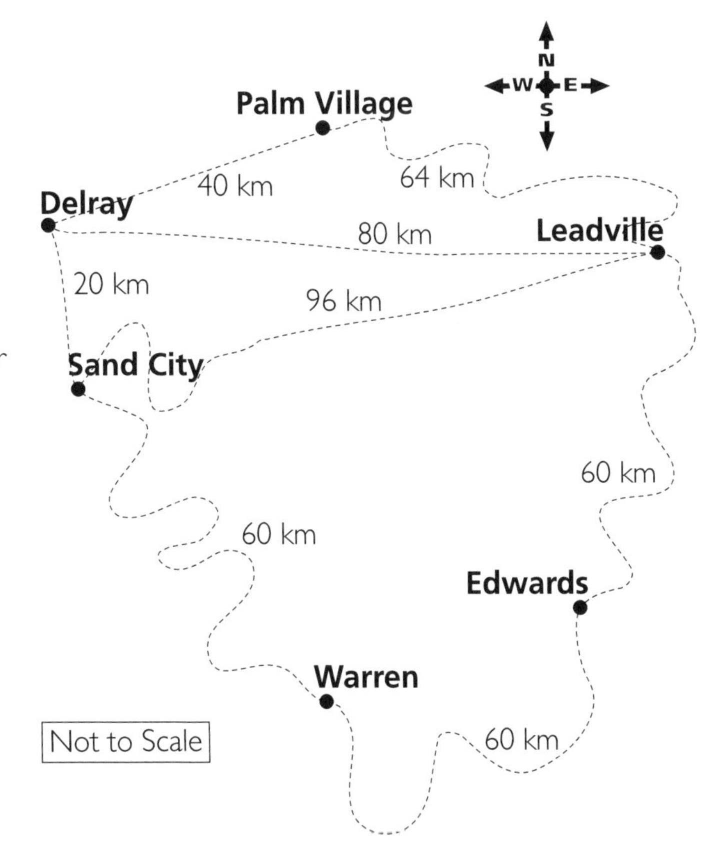

In Scale (F)

Solutions

Use the ratio $\frac{2 \text{ cm}}{3 \text{ km}}$ and a proportion to compute distances between cities in kilometers. To find what distance 11 centimeters on the map represents in terms of kilometers, use a proportion: $\frac{2 \text{ cm}}{3 \text{ km}} = \frac{11 \text{ cm}}{N \text{ km}}$; $2N = 33$; $N = 16.5$ km; 11 cm represents 16.5 km.

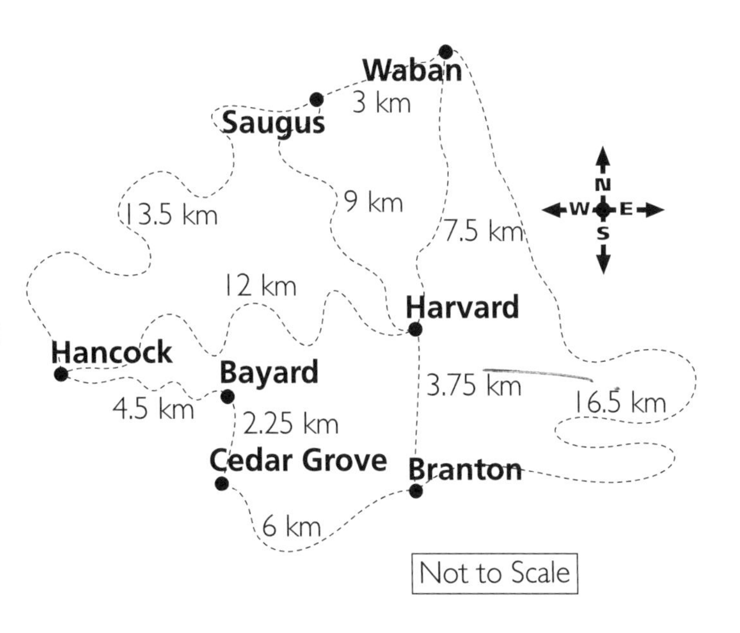

Balance It (A)

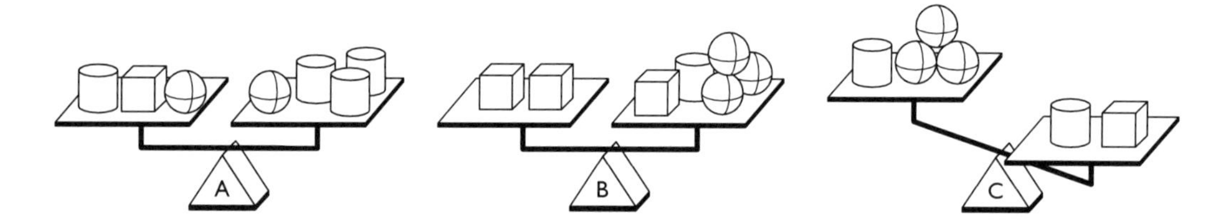

1 Which block, a cylinder, a sphere, or a cube, will balance Scale C?

______

2 List or draw the steps you followed to identify the block.

Name ....

Permission is given by the publisher to the purchasing teacher or parent to reproduce this page for classroom or home use only.

© Creative Publications

Balance It (A)

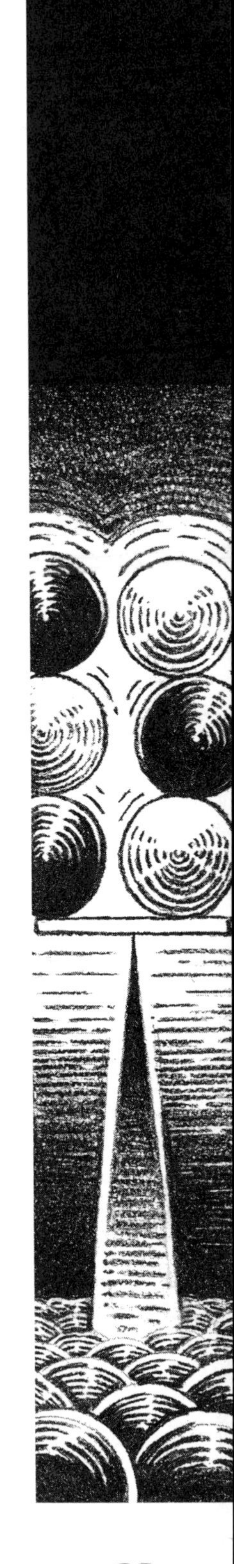

Goals

- Deduce the relationship among mass of objects from visual clues.
- Recognize that balance represents equality.
- Identify collections of objects having equal mass.
- Use substitution as a method of equation solving.

Questions to Ask

- ***On Scale A what blocks are on the left pan that are also on the right pan?*** (One cylinder and one sphere)
- ***On Scale C which pan is heavier?*** (Right pan)
- ***On Scale B if you remove one cube from the left pan and one cube from the right pan, will the scale stay balanced?*** (Yes) ***Why or why not?*** (Removing the same mass from both pans lessens the total mass on each pan by the same amount.)

Solutions

1 Put one cylinder on the left pan.

2 One possible solution method:

a: On Scale A remove one cylinder and one sphere from each pan. That leaves a cube on the left balancing two cylinders on the right. One cube is equal in mass to two cylinders.

b: On Scale B remove one cube from each pan, leaving one cube on the left balancing one cylinder and three spheres on the right.

c: On Scale C substitute one cylinder and three spheres for the cube on the pan on the right. That makes a cylinder and three spheres on the left and two cylinders and three spheres on the right.

d: Add one cylinder to the left pan.

© Creative Publications

Balance It (B)

Name

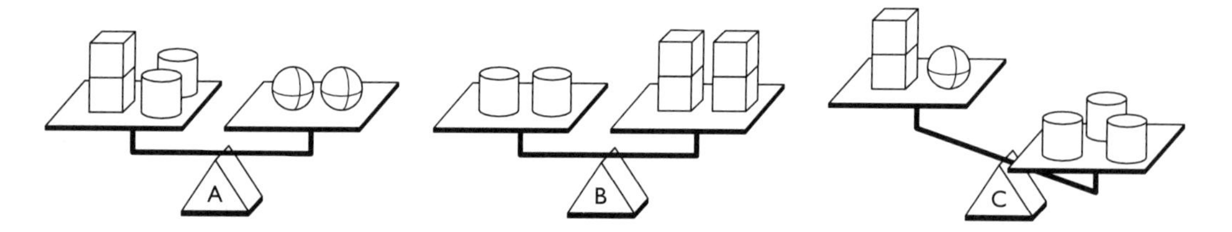

1 Which block, a cylinder, a sphere, or a cube will balance Scale C? _______

2 List or draw the steps you followed to identify the block.

Permission is given by the publisher to the purchasing teacher or parent to reproduce this page for classroom or home use only.

Balance It (C)

Name

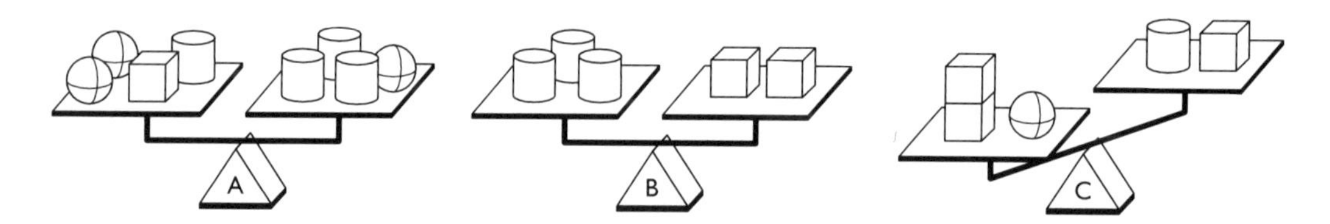

1 Which block, a cylinder, a sphere, or a cube will balance Scale C? _______

2 List or draw the steps you followed to identify the block.

Permission is given by the publisher to the purchasing teacher or parent to reproduce this page for classroom or home use only.

© Creative Publications

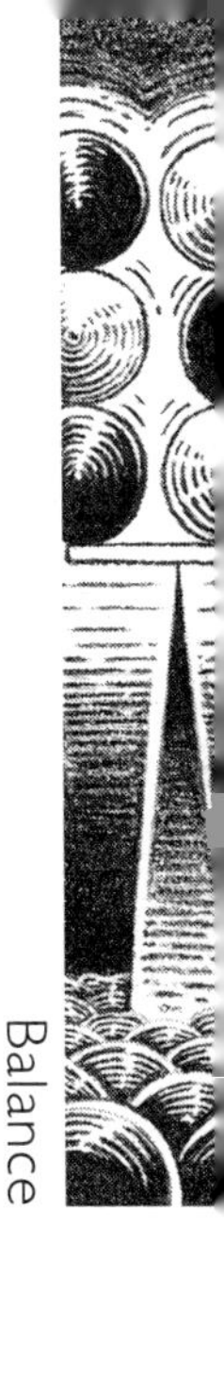

Balance It (B)

Solutions

1 Put one cube on the left pan.

2 One possible solution method:

a: Since two cylinders on Scale B balance four cubes, then one cylinder equals two cubes in mass.

b: Substitute two cubes on Scale A for each cylinder on the left. That makes six cubes on the left balancing two spheres on the right. One sphere equals three cubes in mass.

c: Substitute three cubes on Scale C for the sphere on the left. Substitute two cubes for each cylinder on the right. That puts five cubes on the left and six cubes on the right.

d: Put one cube on the left pan.

Balance

Balance It (C)

Solutions

1 Put one cylinder on the right pan.

2 One possible solution method:

a: Remove one sphere and one cylinder from each pan of Scale A. That leaves a sphere and a cube balancing two cylinders.

b: Add a sphere to each pan of Scale C. That makes two spheres and two cubes on the left and one sphere, one cube, and one cylinder on the right.

c: Substitute two cylinders for each sphere-and-cube pair on Scale C. That puts four cylinders on the left pan and three cylinders on the right.

d: Put one cylinder on the right pan.

© Creative Publications

Balance It (D)

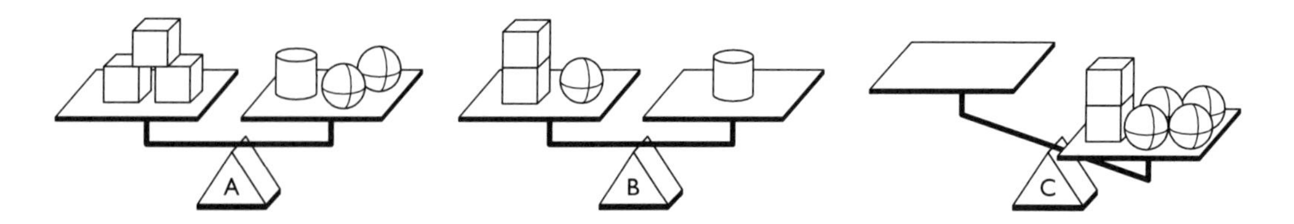

1 Which two blocks will balance Scale C? ______

2 List or draw the steps you followed to solve the problem.

Name ....

Permission is given by the publisher to the purchasing teacher or parent to reproduce this page for classroom or home use only.

© Creative Publications

Balance It (D)

Goals

- Deduce relationships among mass of objects from visual clues.
- Recognize that balance represents equality.
- Identify collections of objects having equal mass.
- Use substitution as a method for equation solving.

Questions to Ask

- *What blocks are on the right pan of Scale C?* (Two cubes and four spheres)
- *What blocks balance the three cubes on Scale A?* (One cylinder and two spheres)
- *On Scale B which block is heavier, the cylinder or a cube?* (The cylinder) *How do you know?* (The cylinder alone balances two cubes and a sphere.)

Solutions

1 Put one cylinder and one cube on the left pan.

2 One possible solution method:

a: On Scale B one cylinder balances two cubes and one sphere. Substitute two cubes and one sphere for the cylinder on Scale A, making two cubes and three spheres on the right balance three cubes on the left.

b: Remove two cubes from each pan of Scale A. That leaves one cube on the left balancing three spheres on the right.

c: Since one cylinder is equal in mass to two cubes and one sphere (see Scale B), put one cylinder on the left pan of Scale C, balancing the two cubes and one of the spheres on the right. There are three spheres on the right that are not yet balanced. Since one cube is equal in mass to three spheres, put a cube on the left pan of Scale C.

d: The two blocks on the left pan, the cylinder and the cube, balance the two cubes and the four spheres on the right pan.

© Creative Publications

Balance It (E)

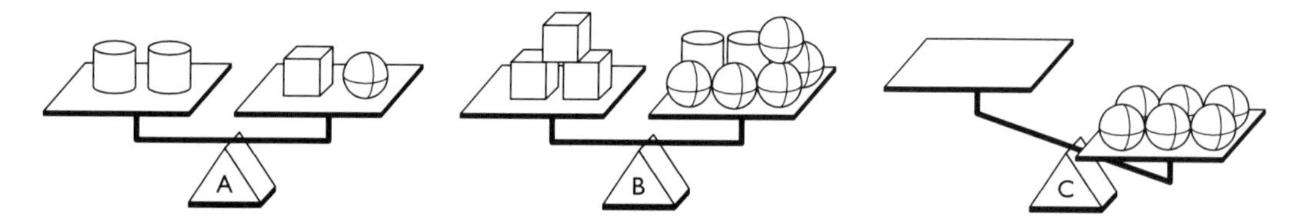

1 Which two blocks will balance Scale C? ______

2 List or draw the steps you followed to solve the problem.

Name ....

Permission is given by the publisher to the purchasing teacher or parent to reproduce this page for classroom or home use only.

Balance It (F)

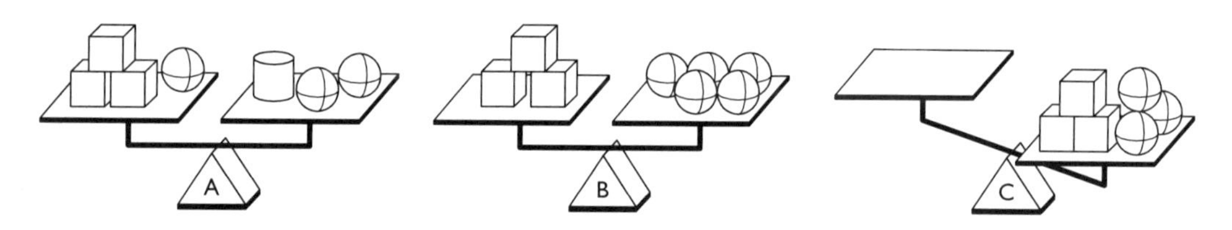

1 Which two blocks will balance Scale C? ______

2 List or draw the steps you followed to solve the problem.

Name ....

Permission is given by the publisher to the purchasing teacher or parent to reproduce this page for classroom or home use only.

© Creative Publications

Balance It (E)

Solutions

1 Put two cubes on the left pan.

2 One possible solution method:

a: Two cylinders balance one cube and one sphere on Scale A.

b: On Scale B substitute one cube and one sphere for the two cylinders on the right pan. That makes one cube and six spheres on the right pan balancing three cubes on the left.

c: Remove one cube from each pan on Scale B. That leaves two cubes on the left balancing six spheres on the right. One cube is equal in mass to three spheres.

d: There are six spheres on the right pan of Scale C. Since three spheres balance one cube, then two cubes balance six spheres. Put two cubes on the left pan.

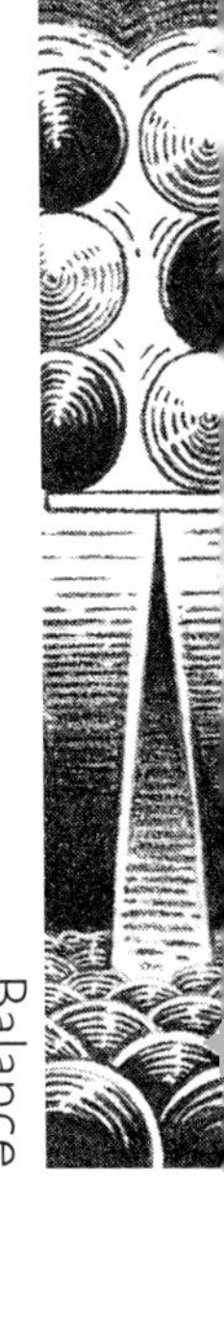

Balance

Balance It (F)

Solutions

1 Put two cylinders on the left pan.

2 One possible solution method:

a: On Scale B three cubes balance five spheres. Substitute five spheres for three cubes on the left pan of Scale A. That makes six spheres on the left pan balancing one cylinder and two spheres on the right.

b: Remove two spheres from each pan of Scale A. That leaves four spheres on the left balancing one cylinder on the right.

c: Put five spheres on the left pan of Scale C to balance the three cubes. Add three more spheres to balance the three spheres on the right. There are now eight spheres in all on the left. Trade them for two cylinders. These two cylinders on the left will balance the scale.

Balance

© Creative Publications

Place It Right (A)

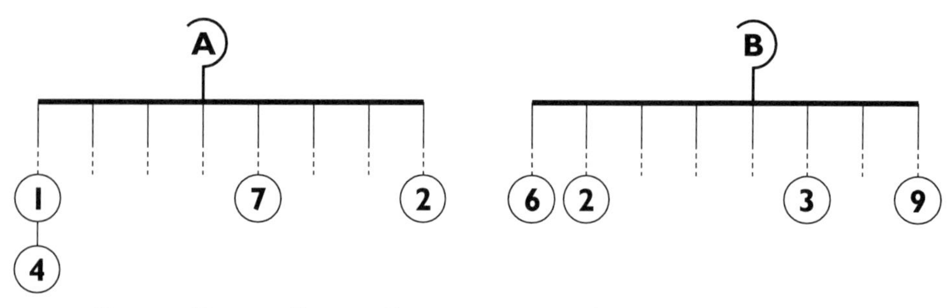

$3 \times (1) + 3 \times (4) = 1 \times (7) + 4 \times (2)$

$3 + 12 = 7 + 8$

$15 = 15$

$4 \times (6) + 3 \times (2) = 1 \times (3) + 3 \times (9)$

$24 + 6 = 3 + 27$

$30 = 30$

The circles represent masses. The number in each circle tells the weight of that mass. The lines on the bars show how far a mass is from the hanger.
These mobiles are balanced.

1 Use all of the numbers given below. Put them in the circles so that the mobile will balance.

2 Describe how you decided where to put the numbers.

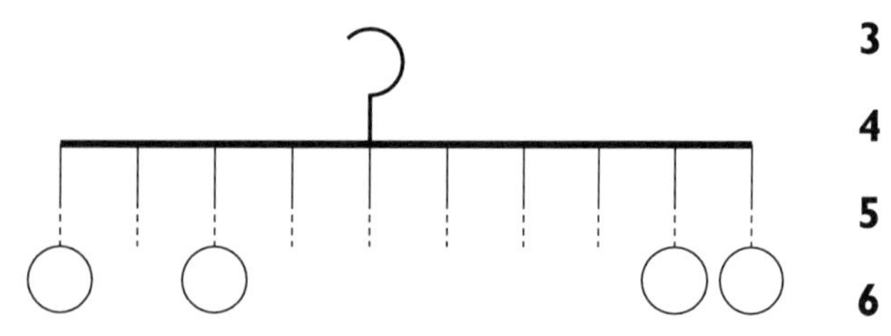

Name ....

Permission is given by the publisher to the purchasing teacher or parent to reproduce this page for classroom or home use only.

© Creative Publications

Place It Right (A)

Goals

- Compute the moment of an object. (The moment is the product of an object's actual mass and its distance from the hanger or balance point.)
- Recognize that equality of moments is represented by a balanced mobile.

Questions to Ask

- *In Mobile A how many masses are hanging on the left side of the hanger?* (2)
- *How many units to the left of the hanger are these masses?* (3 units)
- *In Mobile B how many units to the left of the hanger is the 6-mass?* (4 units)
- *Look at the 9-mass on the right side of Mobile B. What is the product of its distance from the hanger and its mass, or its moment?* (3×9, or 27)

Solutions

1

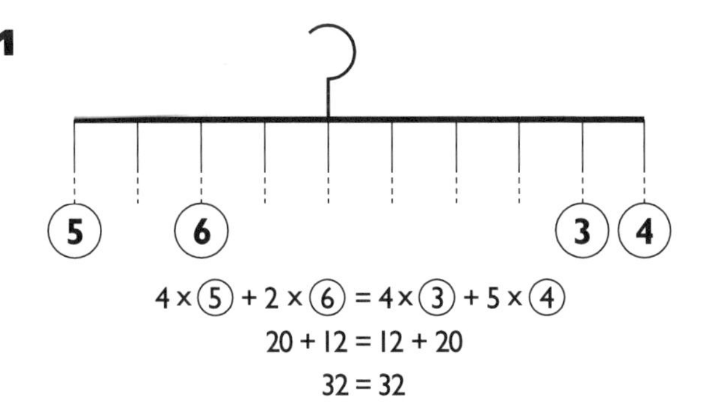

2 One possible solution method:

Write an equation, using M for the masses.

$4 \times M_1 + 2 \times M_2 = 4 \times M_3 + 5 \times M_4$

Look at the masses, hunting for pairs you can put, one on the left and one on the right, that will result in equal products. For example, balance 4×5 on the left with 5×4 on the right. Similarly, 2×6 on the left balances 4×3 on the right.

© Creative Publications

Place It Right (B)

Name

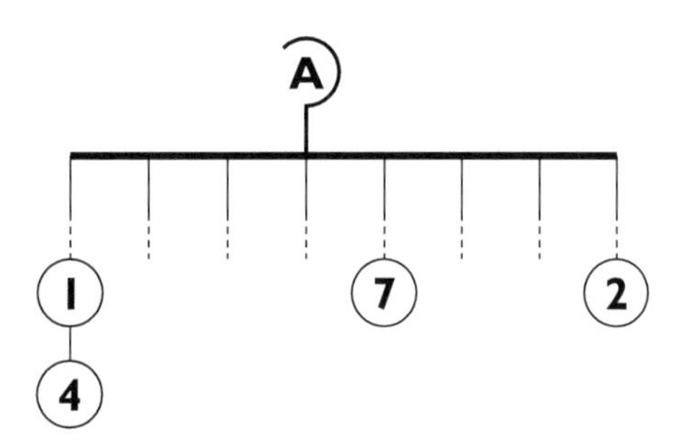

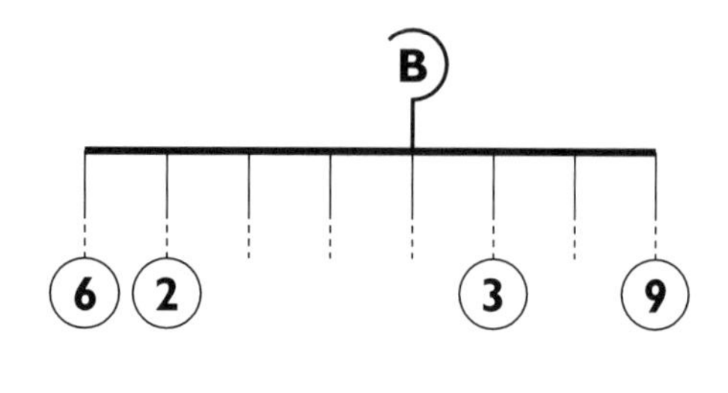

The circles represent masses. The number in each circle tells the weight of that mass. The lines on the bars show how far a mass is from the hanger.
These mobiles are balanced.

1 Use all four of the numbers given. Put them in the circles so that the mobile will balance.

2 Describe how you decided where to put the numbers.

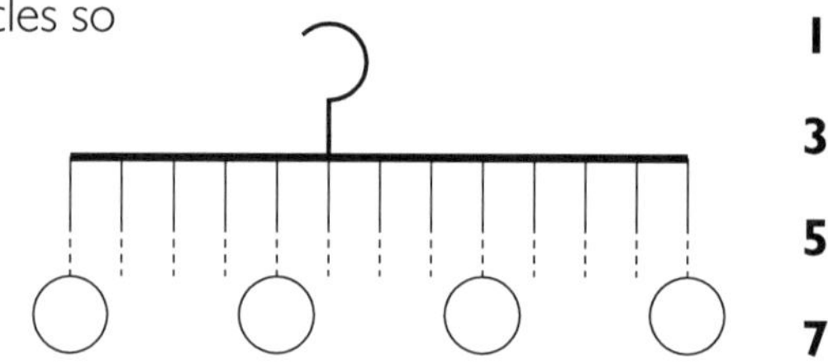

I
3
5
7

Permission is given by the publisher to the purchasing teacher or parent to reproduce this page for classroom or home use only.

Place It Right (C)

Name

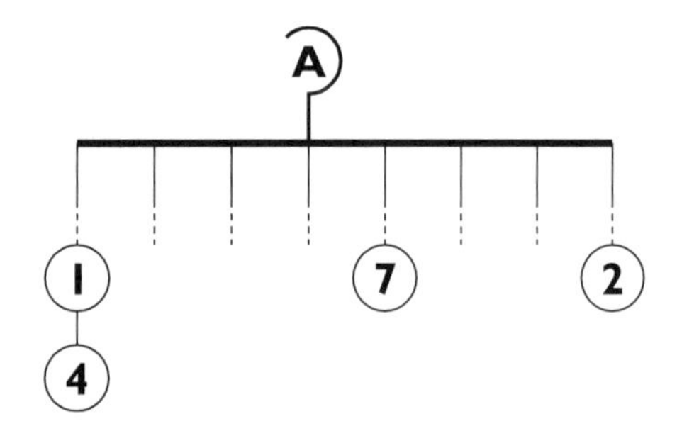

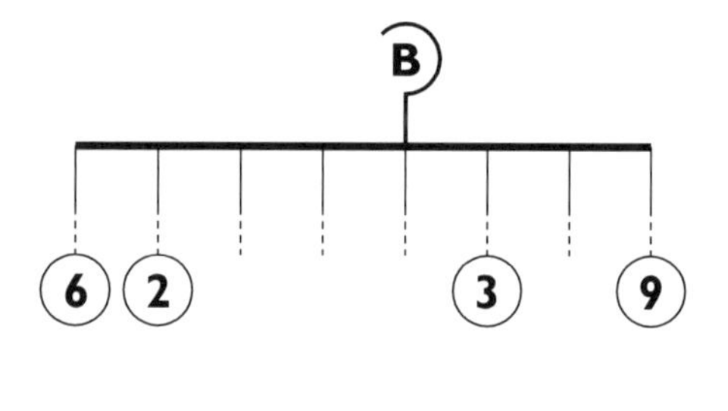

The circles represent masses. The number in each circle tells the weight of that mass. The lines on the bars show how far a mass is from the hanger.
These mobiles are balanced.

1 Use all five of the numbers given. Put them in the circles so that the mobile will balance.

2 Describe how you decided where to put the numbers.

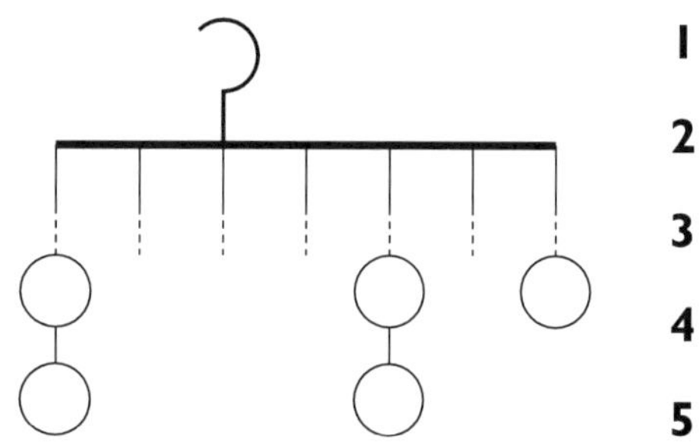

I
2
3
4
5

Permission is given by the publisher to the purchasing teacher or parent to reproduce this page for classroom or home use only.

© Creative Publications

Place It Right (B)

Solutions

1

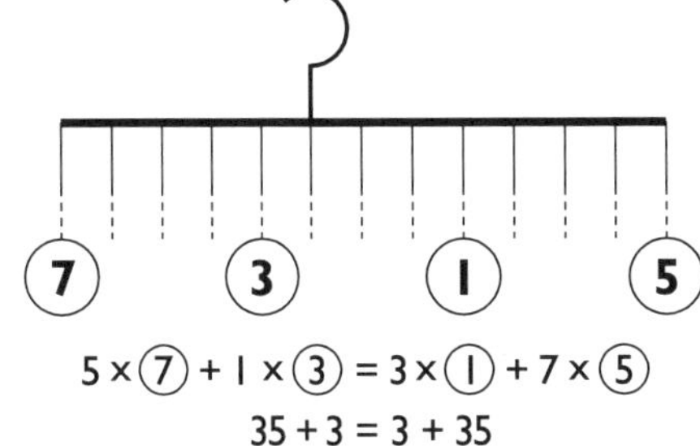

$5 \times (7) + 1 \times (3) = 3 \times (1) + 7 \times (5)$

$35 + 3 = 3 + 35$

$38 = 38$

2 There are at least two ways to place the masses. Here is one possible way:

Write an equation using M for the masses.

$5 \times M_1 + 1 \times M_2 = 3 \times M_3 + 7 \times M_4$

Identify pairs of masses to hang, one on the left and one on the right, to give equal products.

$5 \times 7 = 7 \times 5$; $1 \times 3 = 3 \times 1$

Balance

Place It Right (C)

Solutions

1

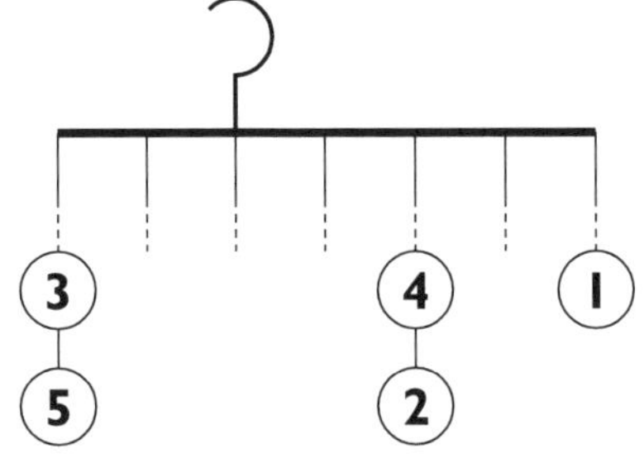

$2 \times (3) + 2 \times (5) = 2 \times (4) + 2 \times (2) + 4 \times (1)$

$6 + 10 = 8 + 4 + 4$

$16 = 16$

2 One possible solution method:

Write an equation, using M for masses.

$2 \times M_1 + 2 \times M_2 = 2 \times M_3 + 2 \times M_4 + 4 \times M_5$

Using the distributive property, rewrite the equation.

$2 \times (M_1 + M_2) = 2 \times (M_3 + M_4) + 4 \times M_5$

Since both sides of the equation involve multiples of 2, divide both sides by 2.

$M_1 + M_2 = M_3 + M_4 + 2 \times M_5$

Think: The sum of two numbers on the left has to be more than the sum of two numbers on the right. The difference between the sums has to be a multiple of 2.

Try different combinations in the equation. The solution is
$5 + 3 = 4 + 2 + 2 \times 1$.

Balance

© Creative Publications

Place It Right (D)

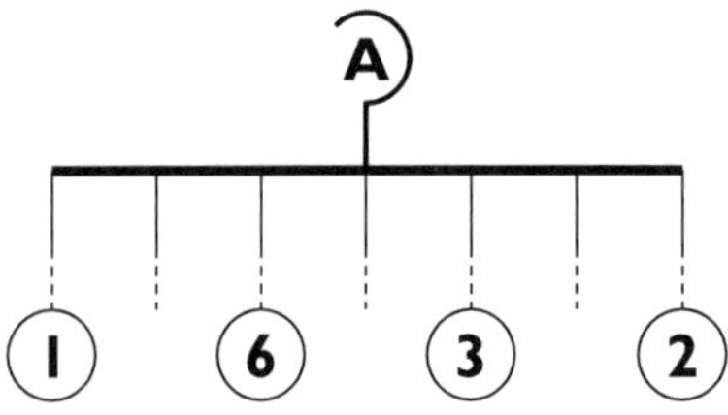

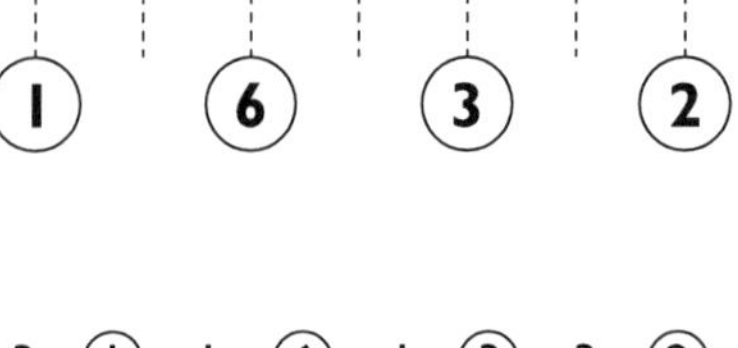

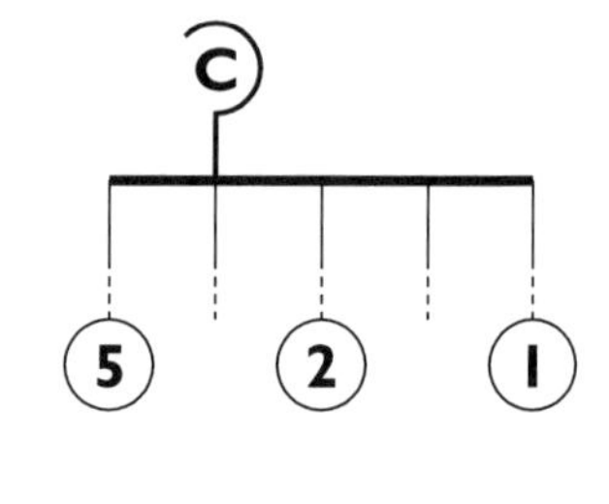

B: 2, 1, 3

$3 \times (1) + 1 \times (6) = 1 \times (3) + 3 \times (2)$

$3 + 6 = 3 + 6$

$9 = 9$

$2 \times [(1) + (2)] = 2 \times (3)$

$2 \times 3 = 6$

$6 = 6$

$1 \times (5) = 1 \times (2) + 3 \times (1)$

$5 = 2 + 3$

$5 = 5$

The circles represent masses. The number in each circle tells the weight of that mass. The lines on the bars show how far a mass is from the hanger.

These mobiles are balanced.

1 Use all of the numbers given below. Put them in the circles so that the mobile will balance.

2 Describe how you decided where to put the numbers.

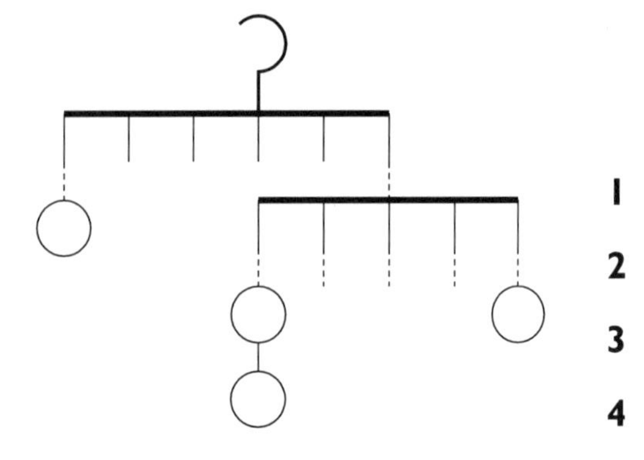

1

2

3

4

Name ....

Permission is given by the publisher to the purchasing teacher or parent to reproduce this page for classroom or home use only.

© Creative Publications

Place It Right (D)

Goals

- Recognize that equality of moments is represented by a balanced mobile.
- Compute the moment of an object. (The moment is the product of its actual mass and its distance from the hanger or balance point.)

Questions to Ask

- ***Why does Mobile A balance?*** (The distance times the mass to the left of the hanger [3 × 1 + 6 × 1, or 9] is equal to the distance times the mass to the right of the hanger [1 × 3 + 3 × 2, or 9].)
- ***On Mobile B the small mobile within the larger one is a mini-mobile. It must balance also. Does it?*** (Yes) ***How?*** (2 × 1 = 1 × 2)
- ***Then what must you do to find the moment of the masses on the left of the hanger?*** (Multiply 3, weight of the mini-mobile, times 2, its distance from the hanger.) ***What is the moment?*** (6)

Solutions

1

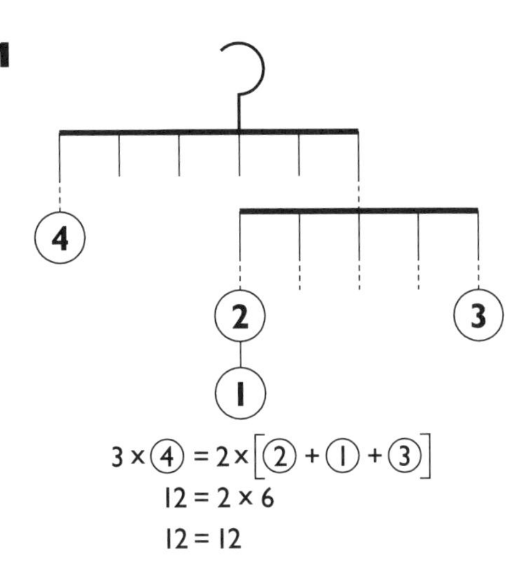

2 One possible solution method:

Students must realize that the moment on the right side of the mobile is 2 times the sum of the masses on the mini-mobile. One mass on the left must balance the mini-mobile. Try the 4-mass on the left. The masses to the right and left of the hanger of the mini-mobile are both two units from its hanger. That means that the sum of the masses on the left must be equal to the mass on the right. The sum must be 3, since that is the weight of the heaviest mass remaining. The masses on the left of the mini-mobile are 2 and 1; the mass on the right is 3.

© Creative Publications

Place It Right (E)

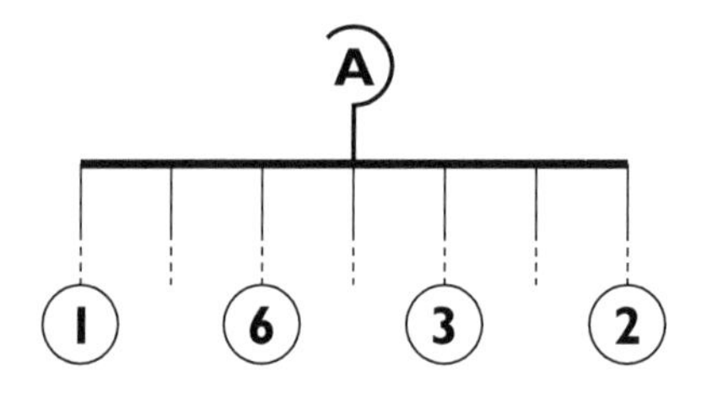

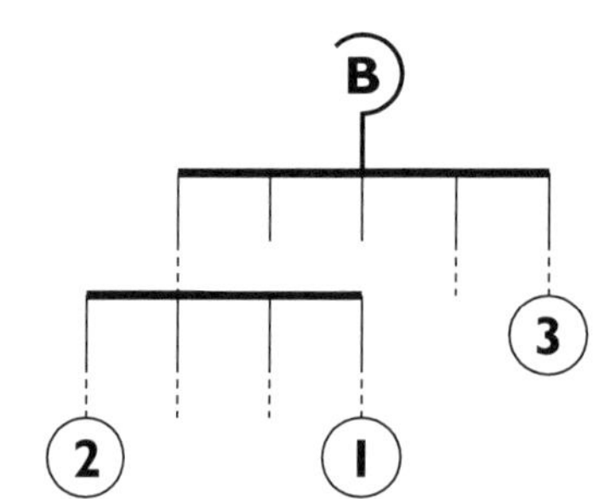

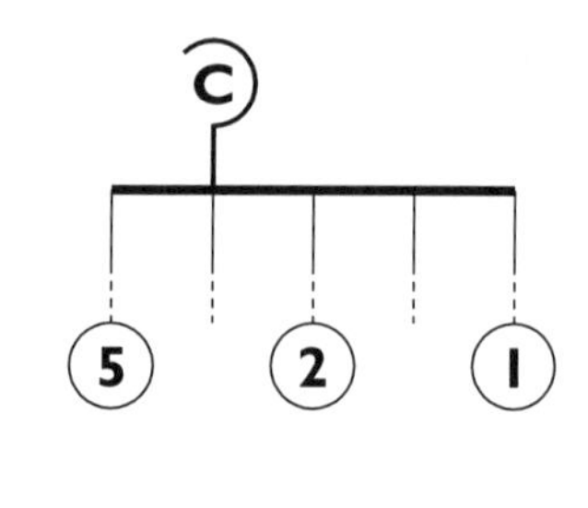

The circles represent masses. The number in each circle tells the weight of that mass. The lines on the bars show how far a mass is from the hanger.
These mobiles are balanced.

1 Use all four of the numbers given. Put them in the circles so that the mobile will balance.

2 Describe how you decided where to put the numbers.

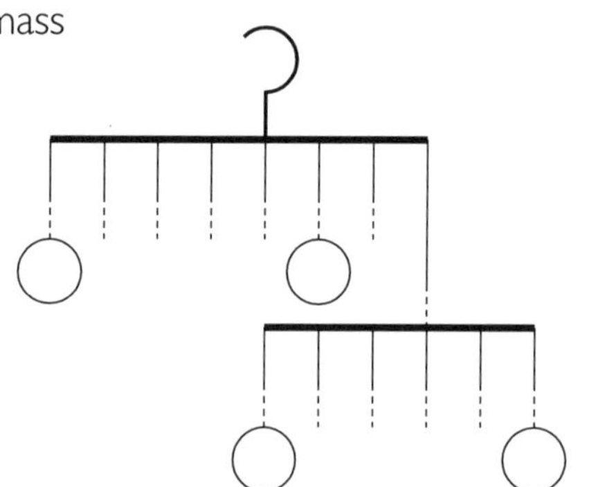

1
2
3
4

Name ....

Permission is given by the publisher to the purchasing teacher or parent to reproduce this page for classroom or home use only.

Place It Right (F)

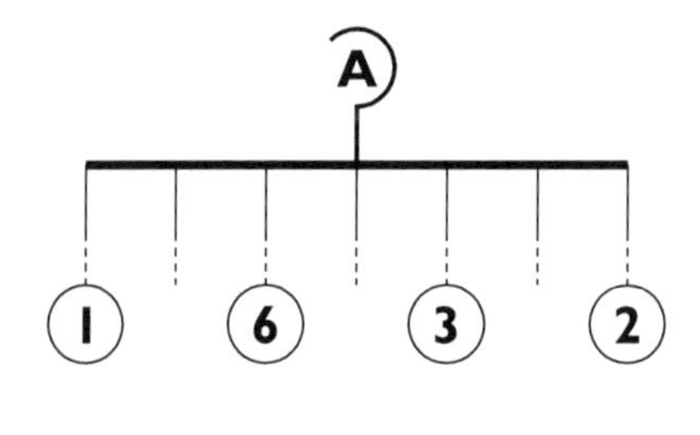

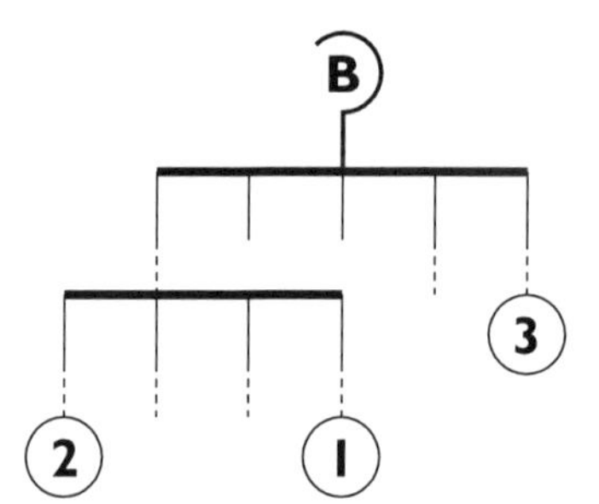

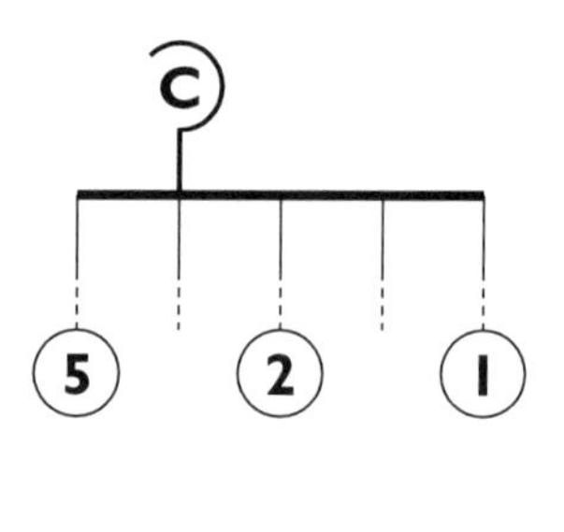

The circles represent masses. The number in each circle tells the weight of that mass. The lines on the bars show how far a mass is from the hanger.
These mobiles are balanced.

1 Use all of the numbers given. Put them in the circles so that the mobile will balance.

2 Describe how you decided where to put the numbers.

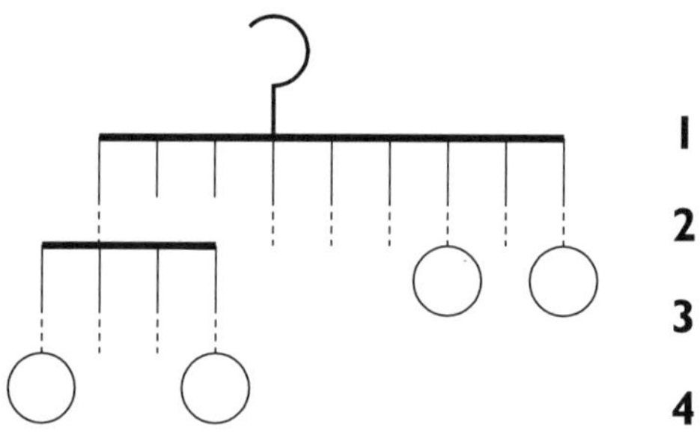

1
2
3
4

Name ....

Permission is given by the publisher to the purchasing teacher or parent to reproduce this page for classroom or home use only.

© Creative Publications

Place It Right (E)

Solutions

1

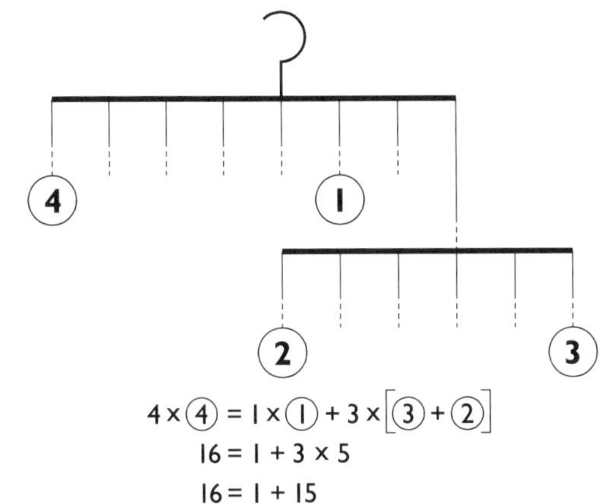

$4 \times (4) = 1 \times (1) + 3 \times [(3) + (2)]$

$16 = 1 + 3 \times 5$

$16 = 1 + 15$

$16 = 16$

2 One possible solution method:

Write an equation. Use M for the masses.

$4 \times M_1 = 1 \times M_2 + 3 \times (M_3 + M_4)$

For the mini-mobile on the right to balance, it must satisfy the equation. The only possible masses to use here are the 2 and 3. From this point use a guess-and-check strategy to find the correct placement of the other masses.

Place It Right (F)

Solutions

1

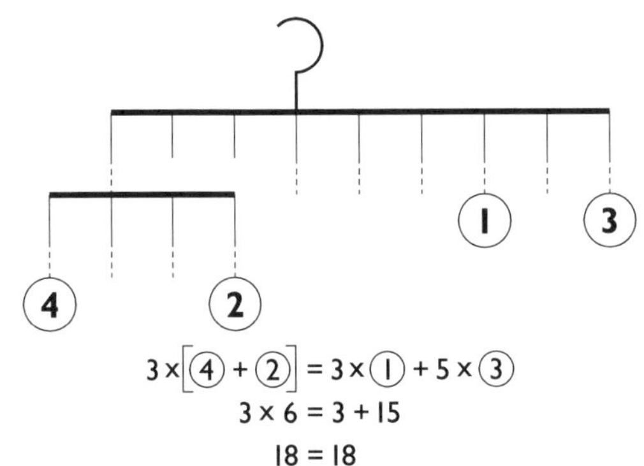

$3 \times [(4) + (2)] = 3 \times (1) + 5 \times (3)$

$3 \times 6 = 3 + 15$

$18 = 18$

2 One possible solution method:

Write an equation. Use M to represent the masses.

For the mini-mobile on the left to balance, it must satisfy the equation, $1 \times M_1 = 2 \times M_2$. The possible sets of masses to use here are 4 and 2 or 2 and 1. Use a guess-and-check strategy to find the correct placement of the other masses.

Balance

Unknown Weights (A)

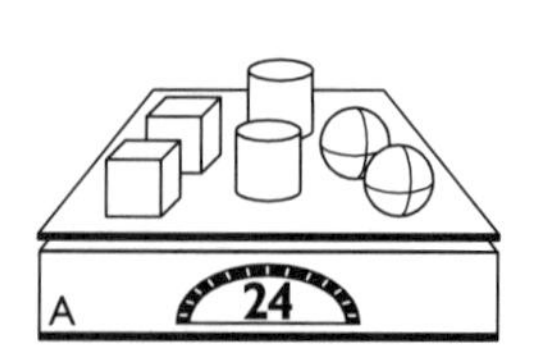

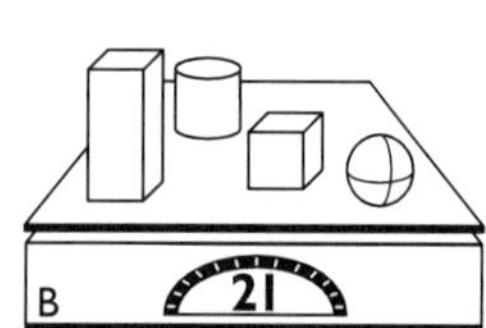

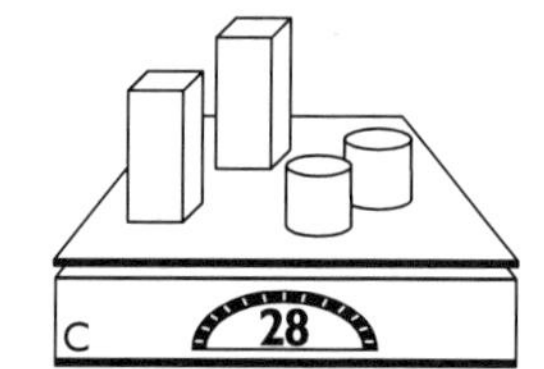

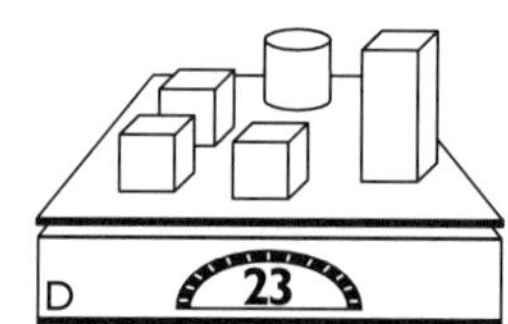

1 Find the weight of each block.
cylinder = ______ pounds
cube = ______ pounds
sphere = ______ pounds
tall prism = ______ pounds

2 Write or draw the steps you followed to solve the problem.

Name ....

Permission is given by the publisher to the purchasing teacher or parent to reproduce this page for classroom or home use only.

© Creative Publications

Unknown Weights (A)

Goals

- Replace symbols with numbers to solve problems.
- Identify values of blocks from relationships shown symbolically.
- Make inferences.

Questions to Ask

- *How many of each type of block are on Scale A?* (2 cubes, 2 spheres, 2 cylinders, 0 tall prisms)
- *What is the total weight on Scale A?* (24 pounds)
- *What is the weight of 1 tall prism and 1 cylinder on Scale C?* (14 pounds) ***How do you know?*** (If all 4 blocks weigh 28 pounds, then half of them weigh 14 pounds.)

Solutions

1 The cylinder weighs 5 pounds.
The cube weighs 3 pounds.
The sphere weighs 4 pounds.
The tall prism weighs 9 pounds.

2 One possible solution method:

a: If the blocks on Scale A weigh 24 pounds, then half of them weigh 12 pounds. The weight of a cylinder, a cube, and a sphere is 12 pounds.

b: If the cube, sphere, and cylinder on Scale B weigh 12 pounds, then the tall prism weighs 9 pounds.

c: If each tall prism on Scale C weighs 9 pounds, the 2 prisms weigh 18 pounds. The 2 cylinders weigh 28 – 18, or 10, pounds. One cylinder weighs 5 pounds.

d: On Scale D if the cylinder weighs 5 pounds and the tall prism weighs 9 pounds, then the 3 cubes weigh 23 – 14, or 9, pounds. Each cube weighs 3 pounds.

e: On Scale B the tall prism weighs 9 pounds, the cube weighs 3 pounds, and the cylinder weighs 5 pounds. The sphere weighs 21 – 9 – 3 – 5, or 4, pounds.

© Creative Publications

Unknown Weights (B)

Name

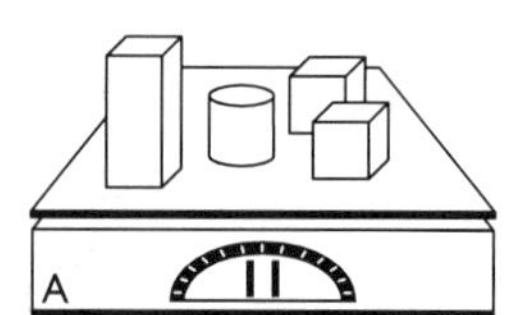

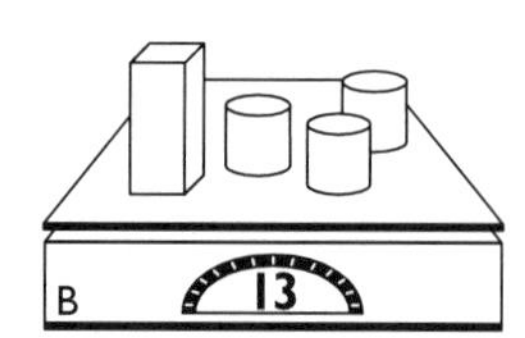

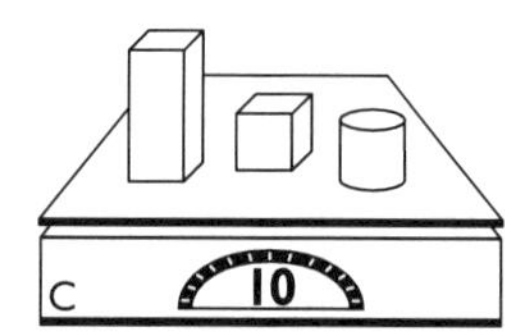

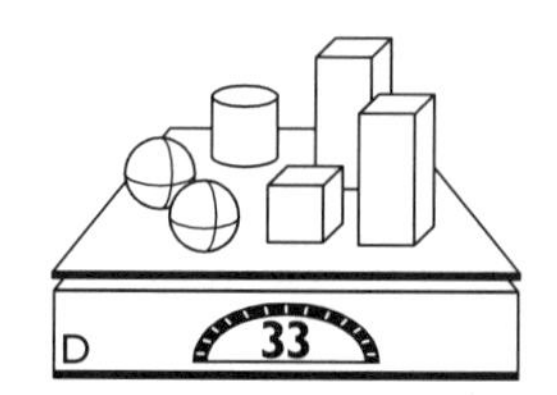

1 Find the weight of each block.
cylinder = _______ pounds
cube = _______ pounds
sphere = _______ pounds
tall prism = _______ pounds

2 Write or draw the steps you followed to solve the problem.

Permission is given by the publisher to the purchasing teacher or parent to reproduce this page for classroom or home use only.

Unknown Weights (C)

Name

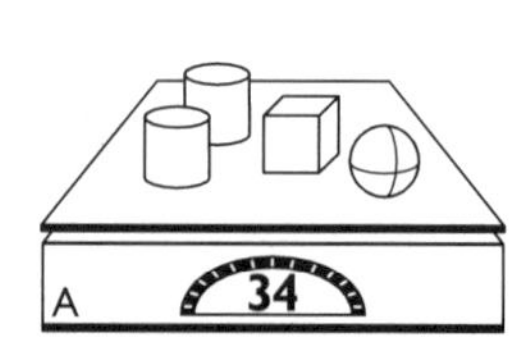

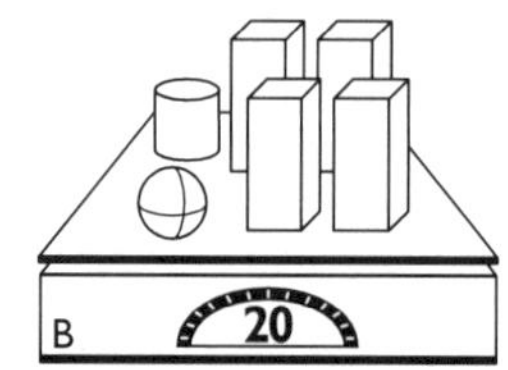

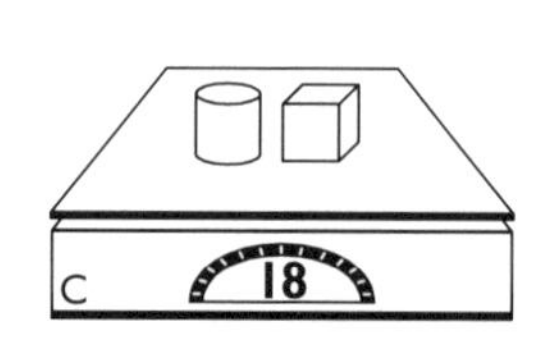

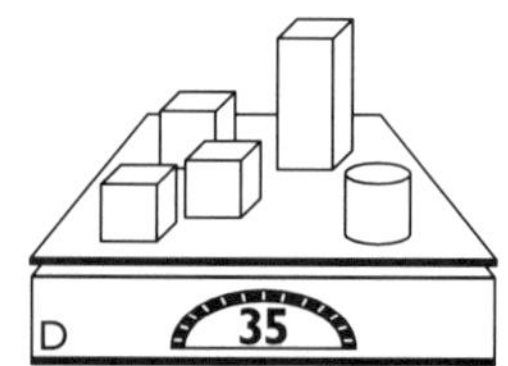

1 Find the weight of each block.
cylinder = _______ pounds
cube = _______ pounds
sphere = _______ pounds
tall prism = _______ pounds

2 Write or draw the steps you followed to solve the problem.

Permission is given by the publisher to the purchasing teacher or parent to reproduce this page for classroom or home use only.

© Creative Publications

Unknown Weights (B)

Solutions

1 The cylinder weighs 2 pounds.
The cube weighs 1 pound.
The sphere weighs 8 pounds.
The tall prism weighs 7 pounds.

2 One possible solution method:

a: On Scale C the blocks weigh a total of 10 pounds. Replace these same blocks on Scale A with 10 pounds. That leaves 1 cube, that would weigh 1 pound.

b: If each cube on Scale A weighs 1 pound, then 2 cubes weigh 2 pounds, and the tall prism and the cylinder weigh 11 – 2, or 9, pounds.

c: On Scale B replace 1 cylinder and the tall prism with 9 pounds. That leaves 2 cylinders that weigh 13 – 9, or 4 pounds. Each cylinder weighs 2 pounds.

d: On Scale C replace the cube with 1 pound and the cylinder with 2 pounds. The tall prism weighs 7 pounds.

e: On Scale D replace the cylinder with 2 pounds, the cube with 1 pound, and each tall prism with 7 pounds. That leaves 2 spheres that weigh 33 – 2 – 1 – 7 – 7, or 16, pounds. Each sphere weighs 8 pounds.

Unknown Weights (C)

Solutions

1 The cylinder weighs 10 pounds.
The cube weighs 8 pounds.
The sphere weighs 6 pounds.
The tall prism weighs 1 pound.

2 One possible solution method:

a: On Scale C a cylinder and a cube weigh 18 pounds. Replace 1 cylinder and the cube on Scale A with 18 pounds, leaving 1 cylinder and 1 sphere weighing 16 pounds.

b: Replace the cylinder and the sphere on Scale B with 16 pounds. The 4 tall prisms weigh 4 pounds; each prism weighs 1 pound.

c: Replace the cylinder and 1 cube on Scale D with 18 pounds and the tall prism with 1 pound. Two cubes weigh 35 – 18 – 1, or 16, pounds; each cube weighs 8 pounds.

d: If the cube on Scale C weighs 8 pounds, then the cylinder weighs 10 pounds.

e: On Scale A replace each cylinder with 10 pounds and the cube with 8 pounds. That leaves the sphere which weighs 34 – 10 – 10 – 8, or 6, pounds.

© Creative Publications

Unknown Weights (D)

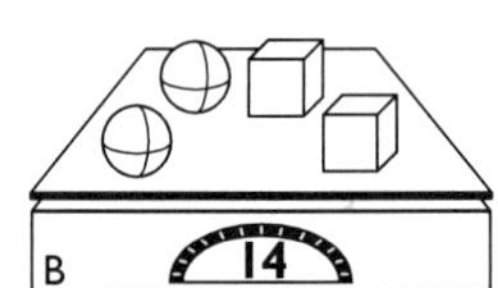

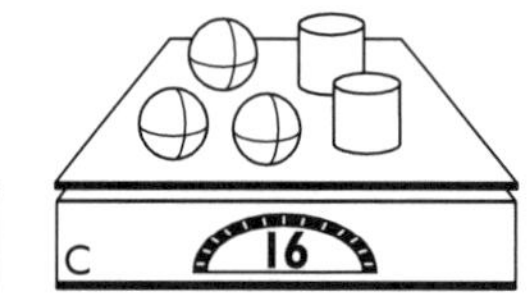

1 Find the weight of each block.
cylinder = ______ pounds
cube = ______ pounds
sphere = ______ pounds
cone = ______ pounds

2 Write or draw the steps you followed to solve the problem.

Name ....

Permission is given by the publisher to the purchasing teacher or parent to reproduce this page for classroom or home use only.

© Creative Publications

Unknown Weights (D)

Goals

- Identify values of blocks from relationships shown symbolically.
- Replace symbols with numbers to solve problems.
- Make inferences.

Questions to Ask

- *If each cylinder on Scale A weighs 1 pound, what does the cone weigh?* (7 pounds) *How do you know?* (Since 2 cylinders and 1 cone weigh 9 pounds, if each cylinder weighs 1 pound, then the cone weighs 9 minus 2, or 7, pounds.)
- *What is the greatest whole number amount that 1 cylinder could weigh?* (4 pounds) *Why?* (Two cylinders and 1 cone on Scale A weigh 9 pounds. If each cylinder weighed more than 4 pounds, then the total would be more than 9 pounds.)
- *How much do 1 cube and 1 sphere weigh?* (7 pounds) *How do you know?* (Scale B shows 2 cubes and 2 spheres weigh 14 pounds, so the weight of 1 of each would be 7 pounds.)

Solutions

1 The cylinder weighs 2 pounds.
The cube weighs 3 pounds.
The sphere weighs 4 pounds.
The cone weighs 5 pounds.

2 One possible solution method:

a: All the blocks on Scale A are also on Scale D. Removing those 3 blocks from Scale D leaves 2 cones that weigh $19 - 9$, or 10, pounds. Each cone weighs 5 pounds.

b: If each cone on Scale D weighs 5 pounds, then 3 cones weigh 15 pounds. That leaves 4 pounds for the 2 cylinders; each cylinder weighs 2 pounds.

c: Each cylinder on Scale C weighs 2 pounds, so the 3 spheres weigh $16 - (2 + 2)$, or 12, pounds. Each sphere weighs $12 \div 3$, or 4, pounds.

d: Two spheres and 2 cubes on Scale B weigh 14 pounds. Each sphere weighs 4 pounds, so the cubes weigh $14 - (4 + 4)$, or 6, pounds. Each cube weighs 3 pounds.

© Creative Publications

Unknown Weights (E)

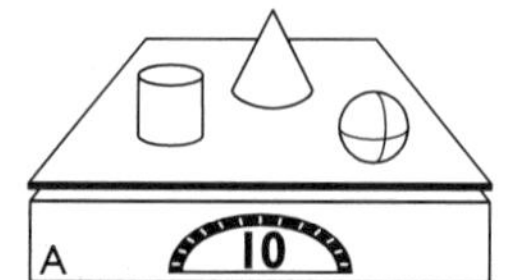

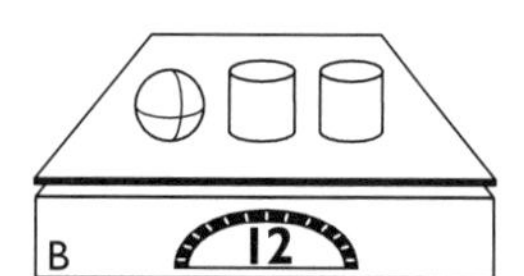

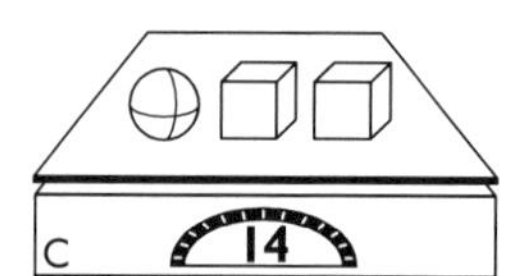

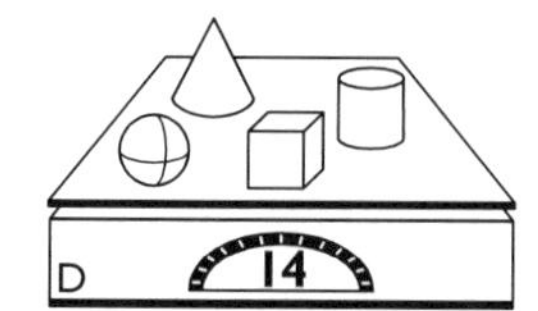

1 Find the weight of each block.
cylinder = _______ pounds
cube = _______ pounds
sphere = _______ pounds
cone = _______ pounds

2 Write or draw the steps you followed to solve the problem.

Name ....

Permission is given by the publisher to the purchasing teacher or parent to reproduce this page for classroom or home use only.

Unknown Weights (F)

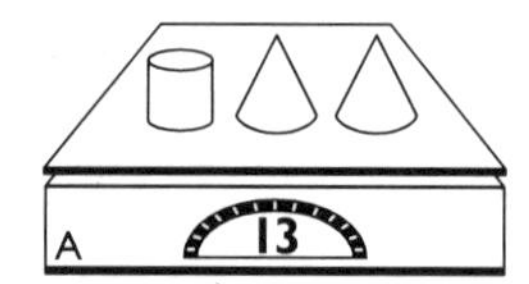

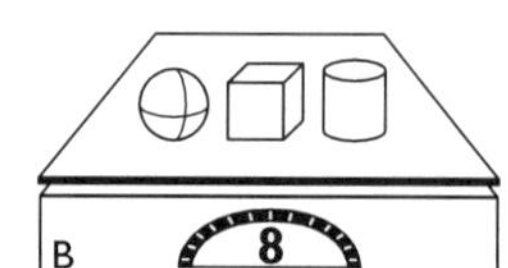

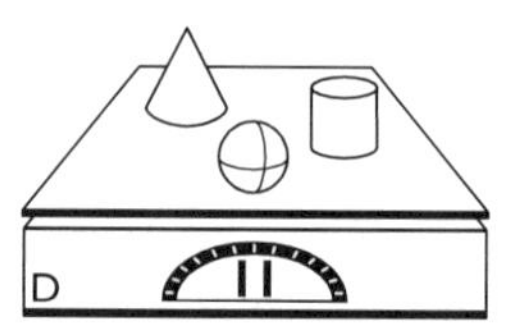

1 Find the weight of each block.
cylinder = _______ pounds
cube = _______ pounds
sphere = _______ pounds
cone = _______ pounds

2 Write or draw the steps you followed to solve the problem.

Name ....

Permission is given by the publisher to the purchasing teacher or parent to reproduce this page for classroom or home use only.

© Creative Publications

Unknown Weights (E)

Solutions

1 The cylinder weighs 3 pounds.
The cube weighs 4 pounds.
The sphere weighs 6 pounds.
The cone weighs 1 pound.

2 One possible solution method:

a: Scale D shows 1 of each kind of block. Scale A shows all of the blocks except the cube. If the blocks on Scale A weigh 10 pounds, then the cube must weigh 4 pounds.

b: If each cube on Scale C weighs 4 pounds, then the sphere weighs 6 pounds.

c: If the sphere on Scale B weighs 6 pounds, then both cylinders also weigh 6 pounds. One cylinder weighs 3 pounds.

d: The cylinder and sphere on Scale A weigh 9 pounds. So the cone weighs 1 pound.

Variable

Unknown Weights (F)

Solutions

1 The cylinder weighs 5 pounds.
The cube weighs 1 pound.
The sphere weighs 2 pounds.
The cone weighs 4 pounds.

2 One possible solution method:

a: Two cylinders and 2 spheres on Scale C weigh 14 pounds. Therefore 1 cylinder and 1 cube weigh 7 pounds.

b: The cylinder and the sphere on Scale D weigh 7 pounds, so the cone weighs 4 pounds.

c: The cylinder and the sphere on Scale B weigh 7 pounds, so the cube weighs 1 pound.

d: One cone weighs 4 pounds, so the 2 cones on Scale A weigh 8 pounds, and the cylinder weighs 5 pounds.

e: If each cylinder on Scale C weighs 5 pounds, 2 of them weigh 10 pounds, and the 2 spheres weigh 4 pounds. Each sphere weighs 2 pounds.

Variable

Solve It (A)

Hexagon + Square = 12

Square + Triangle + Hexagon = 22

Triangle − Hexagon = 1

Same shapes have same numbers.
Different shapes have different numbers.

1 Write numbers in the shapes to make the equations true.

2 Describe how you found the number for Square, for Triangle, and for Hexagon.

Name ....

Permission is given by the publisher to the purchasing teacher or parent to reproduce this page for classroom or home use only.

© Creative Publications

Solve It (A)

Goals

- Replace variables with numbers in systems of equations.
- Identify relationships among variables.
- Use substitution as a method for solving systems of equations.
- Make inferences.

Questions to Ask

- ***What shapes are in the top equation?*** (Hexagon and Square)
- ***In the top equation, is it possible for Hexagon to be 6?*** (No) ***Why or why not?*** (If Hexagon is 6, then Square has to be 6, too. Different shapes must have different number values.)
- ***How are the top and the middle equations alike?*** (Both have a hexagon and a square.)
- ***Which is greater, the value of Triangle or the value of Hexagon?*** (Triangle) ***How do you know?*** (In the bottom equation the difference between their values is 1. This means that Triangle is 1 more than Hexagon.)

Solutions

1 Square is 3.
Triangle is 10.
Hexagon is 9.

2 One possible solution method:

Hexagon and Square together have a value of 12.
Replace Hexagon and Square in the middle equation with 12. That leaves Triangle with a value of 22 – 12, or 10.
In the bottom equation, if Triangle is 10, Hexagon is 9.
In the top equation, if Hexagon is 9, then Square is 3.

© Creative Publications

Solve It (B)

⬡ + ⬡ + □ + △ = 57

△ + □ + □ + △ = 34

△ − □ = 1

Same shapes have same numbers.
Different shapes have different numbers.

1 Write numbers in the shapes to make the equations true.

2 Describe how you found the number for Square, for Triangle, and for Hexagon.

Name ....

Permission is given by the publisher to the purchasing teacher or parent to reproduce this page for classroom or home use only.

Solve It (C)

□ + ⬡ + □ + ⬡ + △ = 38

□ − △ = 7

□ + △ + △ + △ = 23

Same shapes have same numbers.
Different shapes have different numbers.

1 Write numbers in the shapes to make the equations true.

2 Describe how you found the number for Square, for Triangle, and for Hexagon.

Name ....

Permission is given by the publisher to the purchasing teacher or parent to reproduce this page for classroom or home use only.

© Creative Publications

Solve It (B)

Solutions

1 Square is 8.
Triangle is 9.
Hexagon is 20.

2 One possible solution method:

In the middle equation 2 triangles and 2 squares total 34. Therefore, Triangle and Square total 17.
In the top equation replace Square and Triangle with 17. That leaves 2 hexagons with a combined value of 57 – 17, or 40. Thus, Hexagon is 20.
In the bottom equation the difference between Triangle and Square is 1: Triangle – Square = 1.

Since Triangle is 1 more than Square, substitute Triangle for Square in the top equation and increase their sum to 18. The value of Triangle is 9. Put 9 in the middle equation and solve for Square, which is 8.

Solve It (C)

Solutions

1 Square is 11.
Triangle is 4.
Hexagon is 6.

2 One possible solution method:

In the middle equation, Triangle is 7 less than Square.
Substitute Triangle for Square in the bottom equation and decrease the sum by 7 to 16. The new equation is 4 Triangles = 16; the value of Triangle is 4.
Replace each triangle in the bottom equation with 4. Square then has a value of 23 – 4 – 4 – 4, or 11.

In the top equation replace each square with 11 and Triangle with 4. Two hexagons have a combined value of 38 – 11 – 11 – 4, or 12. Hexagon has a value of 6.

© Creative Publications

Solve It (D)

□ + □ + ⬡ + ⬡ = 22

□ + △ = 12

□ − ⬡ = 3

Same shapes have same numbers.
Different shapes have different numbers.

1 Write numbers in the shapes to make the equations true.

2 Describe how you found the number for Square, for Triangle, and for Hexagon.

Name ....

Permission is given by the publisher to the purchasing teacher or parent to reproduce this page for classroom or home use only.

© Creative Publications

Solve It (D)

Goals

- Replace variables with numbers in systems of equations.
- Identify relationships among variables.
- Use substitution as a method for solving systems of equations.
- Make inferences.

Questions to Ask

- *How many different shapes are in the top equation?* (2)
- *What is the sum of the numbers in Square and in Hexagon?* (11) *How do you know?* (The sum of the numbers in 2 squares and 2 hexagons is 22, so the sum of the numbers in 1 of each will be 22 ÷ 2, or 11.)
- *If Square is 3, what number would Triangle be in the middle equation?* (12 – 3, or 9)

Solutions

1 Square is 7.
Triangle is 5.
Hexagon is 4.

2 One possible solution method:

In the top equation Square and Hexagon equal 22 ÷ 2, or 11.
The bottom equation shows that Square is 3 more than Hexagon.
Square is 3 more than Hexagon, and the sum of Square plus Hexagon is 11. The sum of 2 squares instead of Hexagon + Square would be 11 + 3, or 14. Square would be 7.
If Square is 7, then Hexagon is 4.
In the middle equation, if Square is 7, then Triangle is 5.

© Creative Publications

Solve It (E)

□ + △ + ⬡ = 11

□ + □ + ⬡ + △ = 13

⬡ + △ + △ = 14

Same shapes have same numbers.
Different shapes have different numbers.

1 Write numbers in the shapes to make the equations true.

2 Describe how you found the number for Square, for Triangle, and for Hexagon.

Name ....

Permission is given by the publisher to the purchasing teacher or parent to reproduce this page for classroom or home use only.

Solve It (F)

□ + □ + △ = 47

⬡ + ⬡ + △ + △ = 58

⬡ + □ + △ = 44

Same shapes have same numbers.
Different shapes have different numbers.

1 Write numbers in the shapes to make the equations true.

2 Describe how you found the number for Square, for Triangle, and for Hexagon.

Name ....

Permission is given by the publisher to the purchasing teacher or parent to reproduce this page for classroom or home use only.

© Creative Publications

Solve It (E)

Solutions

1 Square is 2.
Triangle is 5.
Hexagon is 4.

2 One possible solution method:

The 3 shapes in the top equation are also in the middle equation. Replace 1 square, Triangle, and Hexagon in the middle equation with 11. That leaves Square equal to 13 – 11, or 2.
If Square is 2, then the sum of Triangle and Hexagon is 11 – 2, or 9.
In the bottom equation replace 1 triangle and Hexagon with 9. That leaves Triangle worth 5.
If the sum of Triangle and Hexagon is 9, and Triangle is 5, then Hexagon is 4.

Solve It (F)

Solutions

1 Square is 15.
Triangle is 17.
Hexagon is 12.

2 One possible solution method:

In the middle equation 2 triangles and 2 hexagons total 58, so Triangle + Hexagon is 58 ÷ 2, or 29.
Replace Triangle and Hexagon in the bottom equation with 29. That leaves Square equal to 44 – 29, or 15.
In the top equation, if Square is 15, then Triangle is 47 – (15 + 15), or 17.
In the bottom equation, if Square is 15 and Triangle is 17, then Hexagon is 44 – (15 + 17), or 12.

© Creative Publications

Puzzle Grid (A)

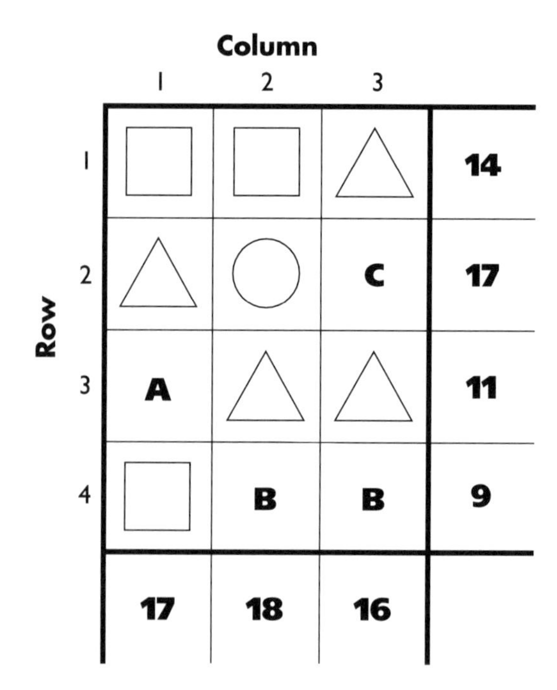

This is a puzzle grid. Write numbers in the shapes to make the equations true.

Same shapes must have same values.
Different shapes must have different values.
The number at the end of each row and each column is the sum.

1 $A =$ ______

2 $B =$ ______

3 $C =$ ______

4 Explain your answers.

Name ....

Permission is given by the publisher to the purchasing teacher or parent to reproduce this page for classroom or home use only.

© Creative Publications

Puzzle Grid (A)

Goals

- Replace variables with numbers to make equations true.
- Identify relationships among variables.
- Use substitution as a method for solving systems of equations.
- Make inferences.

Questions to Ask

- ***Look at the rows and the columns. Is there a column that contains all of the shapes that appear in a row?*** (Yes) ***Which one?*** (Column 1 has the same shapes as Row 1.)
- ***If you replaced Triangle in Row 3 with the number 2, what would the value of A be?*** ($11 - [2 + 2]$, or 7)
- ***If the value of Square in Row 4 were 1, what would the value of B be?*** (Two B's would be $9 - 1$, or 8, and B would be $8 \div 2$, or 4.)

Solutions

1 $A = 3$

2 $B = 2$

3 $C = 6$

4 One possible solution method:

Compare the shapes in Row 1 and Column 1. Column 1 has the letter A and all the shapes that are in Row 1. The value of A is the difference between the sums of Column 1 and Row 1. Thus A is $17 - 14$, or 3.
Replace A in Row 3 with 3. The 2 triangles are worth $11 - 3$, or 8. Thus Triangle is $8 \div 2$, or 4.
Replace Triangle in Row 1 with 4. The 2 squares are worth $14 - 4$, or 10, and Square is $10 \div 2$, or 5.
Replace Square in Row 4 with 5. That means that the 2 B's are $9 - 5$, or 4, and B is $4 \div 2$, or 2.
Replace each triangle in Column 3 with 4, and B with 2. That leaves $C = 16 - (4 + 4 + 2)$, or 6.

© Creative Publications

Puzzle Grid (B)

Name ....

This is a puzzle grid. Write numbers in the shapes to make the equations true.

Same shapes must have same values.
Different shapes must have different values.
The number at the end of each row and each column is the sum.

1 $A =$ ______

2 $B =$ ______

3 $C =$ ______

4 Explain your answers.

Row \ Column	1	2	3	
1	**A**	**B**	◯	**16**
2	◯	△	□	**17**
3	□	□	**C**	**20**
4	**B**	△	◯	**14**
	22	**23**	**22**	

Permission is given by the publisher to the purchasing teacher or parent to reproduce this page for classroom or home use only.

Puzzle Grid (C)

Name ....

This is a puzzle grid. Write numbers in the shapes to make the equations true.

Same shapes must have same values.
Different shapes must have different values.
The number at the end of each row and each column is the sum.

1 $A =$ ______

2 $B =$ ______

3 $C =$ ______

4 Explain your answers.

Row \ Column	1	2	3	
1	□	◯	**A**	**20**
2	□	**C**	◯	**19**
3	◯	**A**	△	**22**
4	**B**	□	△	**21**
	27	**29**	**26**	

Permission is given by the publisher to the purchasing teacher or parent to reproduce this page for classroom or home use only.

© Creative Publications

Puzzle Grid (B)

Solutions

1 $A = 9$

2 $B = 3$

3 $C = 8$

4 One possible solution method:

Look for a column that contains some of the shapes or letters that are in a row. Column 1 contains *A*, *B*, and Circle, that are in Row 1. The value of Square is the difference between the sums of Column 1 and Row 1: $22 - 16 = 6$.
Replace Square in Row 3 with 6.
Then *C* is $20 - (6 + 6)$, or 8.
Replace Square in Column 3 with 6, and *C* with 8.
Two circles are $22 - (6 + 8)$, or 8.
Circle is $8 \div 2$, or 4.
Replace Circle in Row 2 with 4 and Square with 6. That means that Triangle is $17 - (6 + 4)$, or 7.
Replace Triangle in Row 4 with 7 and Circle with 4. That leaves *B* equal to $14 - (7 + 4)$, or 3.
Replace *B* in Row 1 with 3, and Circle with 4. That leaves *A* equal to $16 - (3 + 4)$, or 9.

Variable

Puzzle Grid (C)

Solutions

1 $A = 10$

2 $B = 15$

3 $C = 9$

4 One possible solution method:

Column 3 contains the letter and shapes that are in Row 3.
The value of 1 triangle is the difference between the sums of Column 3 and Row 3: $26 - 22 = 4$.
Replace each triangle in Column 3 with 4. That means that Circle and *A* together are $26 - (4 + 4)$, or 18.
In Row 1 Circle and *A* are 18; Square is $20 - 18$, or 2.
Replace Square in Row 4 with 2, and Triangle with 4. *B* is $21 - (2 + 4)$, or 15.
Replace each square in Column 1 with 2 and *B* with 15.
Circle is then $27 - (2 + 2 + 15)$, or 8.
In Row 1 replace Square with 2 and Circle with 8.
That leaves $A = 20 - (2 + 8)$, or 10.
In Row 2 replace Square with 2 and Circle with 8.
That leaves $C = 19 - (2 + 8)$, or 9.

Variable

Puzzle Grid (D)

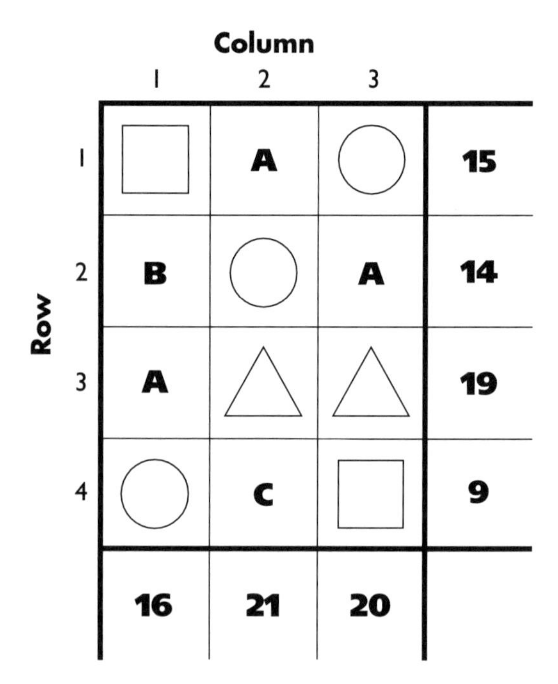

This is a puzzle grid. Write numbers in the shapes to make the equations true.

Same shapes must have same values.
Different shapes must have different values.
The number at the end of each row and each column is the sum.

1 $A =$ ______

2 $B =$ ______

3 $C =$ ______

4 Explain your answers.

Name ....

Permission is given by the publisher to the purchasing teacher or parent to reproduce this page for classroom or home use only.

© Creative Publications

Puzzle Grid (D)

Goals

- Replace variables with numbers to make equations true.
- Identify relationships among variables.
- Use substitution as a method for solving systems of equations.
- Make inferences.

Questions to Ask

- ***How are Rows 1 and 2 alike?*** (They both have an *A* and a circle.) ***How are they different?*** (Row 1 contains a square and Row 2 has the letter *B*. The sum of Row 1 is 1 more than the sum of Row 2.)
- ***Why do you think Row 1 has a greater sum than Row 2?*** (The value of Square must be 1 more than the value of *B*.)
- ***Where would be the best place to start in order to figure out the values in this puzzle?*** (Look for a row and a column that have some of the same letters and shapes.) ***What pair do you see?*** (Row 1 and Column 1)

Solutions

1 $A = 9$

2 $B = 1$

3 $C = 3$

4 One possible solution method:

Square, *A*, and Circle in Row 1 are also in Column 1.
Replace them with 15, the sum of Row 1. Then *B* in Column 1 has a value of 16 – 15, or 1.
If *B* in Row 2 is 1, then Circle and *A* total 13.
If Circle and *A* in Row 1 total 13, then Square is 15 – 13, or 2.
All the variables in Row 1 are in Column 3. Since the value of Row 1 is 15, Triangle in Column 3 equals 20 – 15, or 5.
Replace each triangle in Row 3 with 5. Then *A* equals 19 – 5 – 5, or 9.
In Column 3, Square is 2. That means that the sum of Circle, *A*, and Triangle, is 20 – 2, or 18.
Replace Circle, *A*, and Triangle in Column 2 with 18. That leaves $C = 21 - 18$, or 3.

© Creative Publications

Puzzle Grid (E)

This is a puzzle grid. Write numbers in the shapes to make the equations true.

Same shapes must have same values.
Different shapes must have different values.
The number at the end of each row and each column is the sum.

1 $A =$ ______

2 $B =$ ______

3 $C =$ ______

4 Explain your answers.

Column

Row	1	2	3	
1	○	□	**C**	**17**
2	**C**	**A**	□	**19**
3	**B**	○	□	**16**
4	□	△	**C**	**23**
	24	**23**	**28**	

Name ....

Permission is given by the publisher to the purchasing teacher or parent to reproduce this page for classroom or home use only.

Puzzle Grid (F)

This is a puzzle grid. Write numbers in the shapes to make the equations true.

Same shapes must have same values.
Different shapes must have different values.
The number at the end of each row and each column is the sum.

1 $A =$ ______

2 $B =$ ______

3 $C =$ ______

4 Explain your answers.

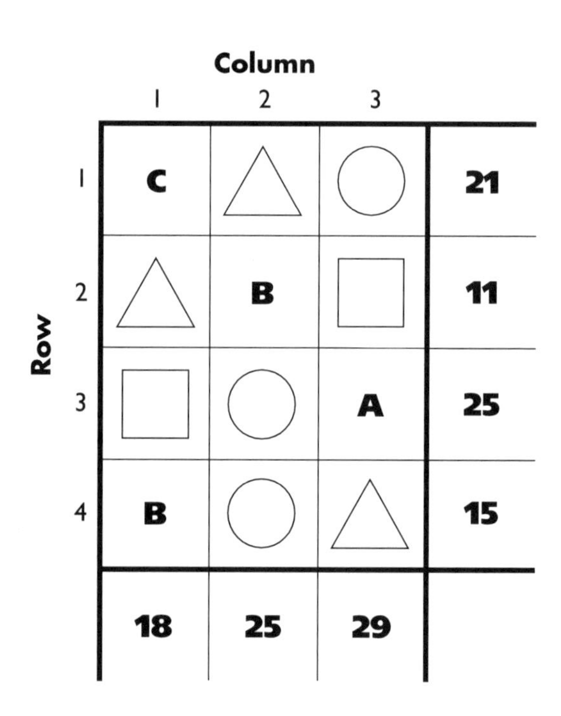

Name ....

Permission is given by the publisher to the purchasing teacher or parent to reproduce this page for classroom or home use only.

© Creative Publications

Puzzle Grid (E)

Solutions

1 $A = 5$

2 $B = 7$

3 $C = 8$

4 One possible solution method:

Look for a row and a column that have some of the same letters and shapes. All of Row 1 is found in Column 1.
Replace Circle, C, and Square in Column 1 with 17. That leaves B equal to $24 - 17$, or 7.
Two squares and 2 C's in Column 3 total 28, so Square + C = 14.
Replace C and Square in Row 2 with 14. That leaves A having a value of $19 - 14$, or 5.
Replace B in Row 3 with 7. That leaves Circle and Square equal to $16 - 7$, or 9.
Replace Circle and Square in Row 1 with 9. That leaves C equal to $17 - 9$, or 8.

Puzzle Grid (F)

Solutions

1 $A = 9$

2 $B = 1$

3 $C = 7$

4 One possible solution method:

Look for a row and a column that have some of the same letters and shapes. Triangle, B, and Square in Row 2 are also in Column 1. Replace them with 11; C then equals 7.
If C in Row 1 is 7, then Triangle and Circle total $21 - 7$, or 14.
Replace Triangle and Circle in Row 4 with 14. That leaves B equal to 1.
Replace B with 1 in Row 2. That leaves Triangle and Square totaling $11 - 1$, or 10.
All of Row 4 is in Column 2. Replace Triangle, B, and 1 circle in Column 2 with 15. That leaves Circle equal to $25 - 15$, or 10.
Replace Triangle and Square in Column 3 with 10, and Circle with 10. Then A equals $29 - 10 - 10$, or 9.

© Creative Publications

Tag Sale (A)

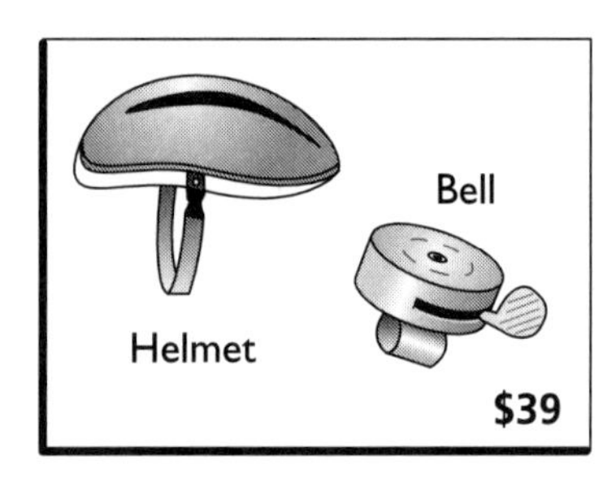

These signs tell about some items for sale. Same items have same prices. Different items have different prices. How much is:

1 A helmet? ______

2 A bell? ______

3 A bicycle lock? ______

4 Explain how you figured out the prices.

Name ....

Permission is given by the publisher to the purchasing teacher or parent to reproduce this page for classroom or home use only.

 © Creative Publications

Tag Sale (A)

Variable

Goals

- Identify relationships among variables.
- Replace variables with numbers in systems of equations.
- Use substitution as a method for solving systems of equations.
- Make inferences.

Questions to Ask

- *How much is a helmet and a bell?* ($39)
- *How are the first two signs alike?* (Both include a bell.)
- *How are the first two signs different?* (The first sign has a helmet, the second has a lock; the prices are different.)
- *What causes the difference in prices between the first two signs?* (The bells are the same price; the helmet must cost more than the lock.)

Solutions

1 The helmet costs $29.

2 The bell costs $10.

3 The bicycle lock costs $24.

4 One possible solution method:

Compare the first two signs. Since the bells cost the same, the difference in prices must be because a helmet costs more than a lock. A helmet costs $5 more than a lock, because the total prices differ by $5.
On the third sign, if both items were helmets, then the total cost would be $5 more than $53, or $58; each helmet would cost $58 ÷ 2, or $29.
On the first sign, if a helmet is $29, then a bell is $10.
On the second sign, if a bell is $10, then a lock is $24.

© Creative Publications

Tag Sale (B)

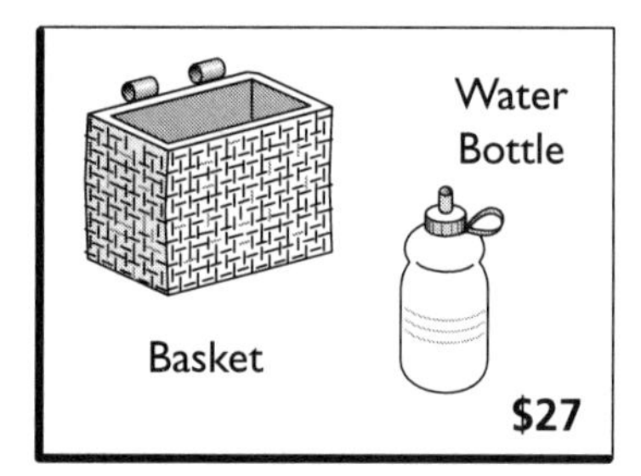

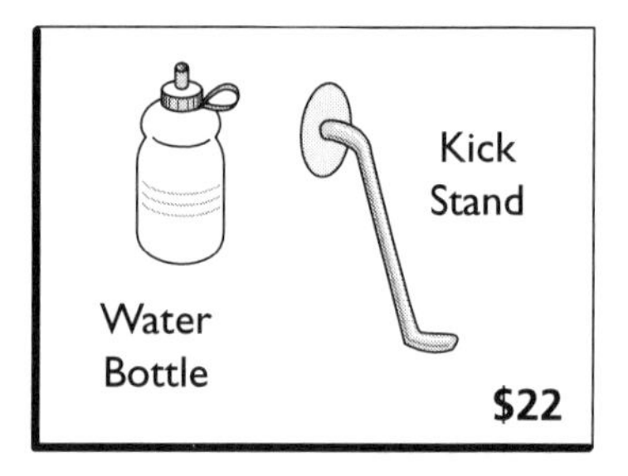

These signs tell about some items for sale. Same items have same prices. Different items have different prices. How much is:

1 A basket? ______

2 A water bottle? ______

3 A kickstand? ______

4 Explain how you figured out the prices.

Name ....

Permission is given by the publisher to the purchasing teacher or parent to reproduce this page for classroom or home use only.

Tag Sale (C)

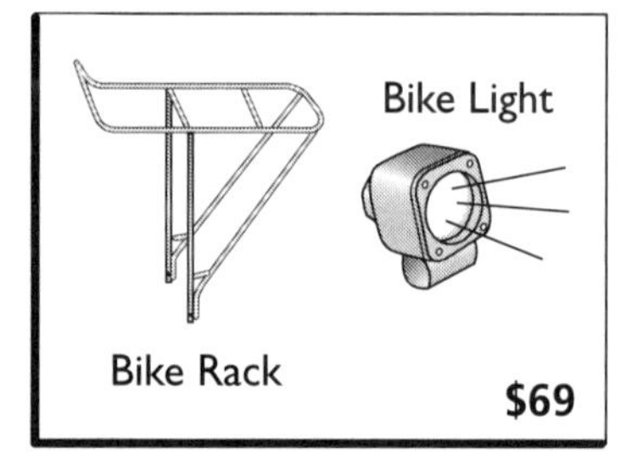

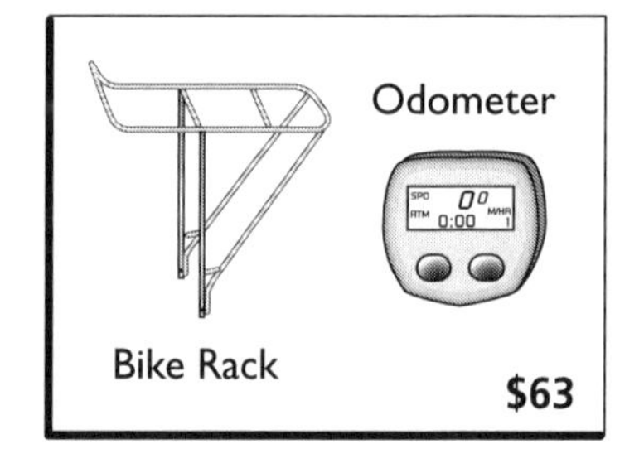

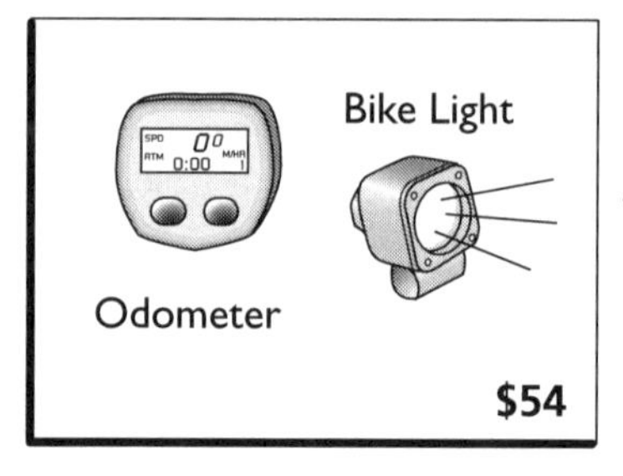

These signs tell about some items for sale. Same items have same prices. Different items have different prices. How much is:

1 A bike light? ______

2 A bike rack? ______

3 An odometer? ______

4 Explain how you figured out the prices.

Name ....

Permission is given by the publisher to the purchasing teacher or parent to reproduce this page for classroom or home use only.

© Creative Publications

Tag Sale (B)

Solutions

1 A basket costs $13.

2 A bottle costs $14.

3 A kickstand costs $8.

4 One possible solution method:

There is a water bottle on each of the first two signs. The other item on the first sign is a basket and on the second sign, a kickstand. The difference in price between the pairs of items, $5, reflects the difference in the prices of a basket and a kickstand. A basket costs $5 more than a kickstand.

If there were another basket instead of a kickstand on the third sign, then the price for two baskets would be $21 + $5, or $26. One basket would cost $13.
On the third sign, if a basket costs $13, then a kickstand is $21 – $13, or $8.
On the second sign, if a kickstand is $8, then the water bottle is $22 – $8, or $14.

Tag Sale (C)

Solutions

1 A bike light costs $30.

2 A bike rack costs $39.

3 An odometer costs $24.

4 One possible solution method:

Compare the first and second signs. A bike light costs $6 more than an odometer.
On the third sign, if there were two lights, then the price would be $6 more, or $60. One light would cost $30.
On the first sign, if a light costs $30, then a bike rack costs $69 – $30, or $39.
On the second sign, if a bike rack costs $39, then an odometer costs $24.

© Creative Publications

Tag Sale (D)

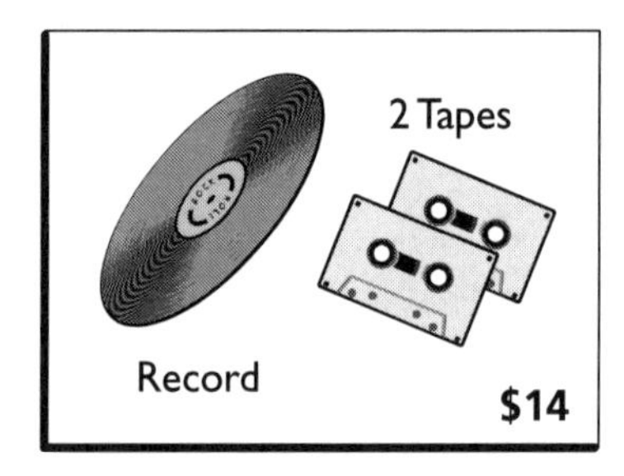

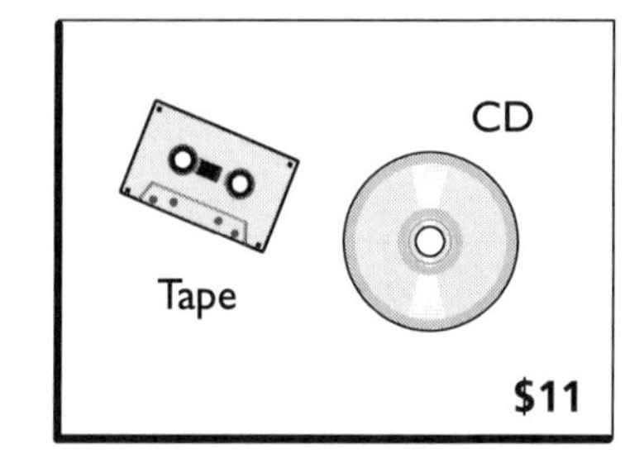

At a tag sale Josh is selling old records, CD's, and tapes. Same items have same prices. Different items have different prices. How much is:

1 One record? ______

2 One CD? ______

3 One tape? ______

4 Explain how you figured out the prices.

Name ....

Permission is given by the publisher to the purchasing teacher or parent to reproduce this page for classroom or home use only.

© Creative Publications

Tag Sale (D)

Goals

- Identify relationships among variables.
- Replace variables with numbers in systems of equations.
- Use substitution as a method for solving systems of equations.
- Make inferences.

Questions to Ask

- ***Compare items on the first and third signs. Which costs more, a record or a tape?*** (A record) ***How do you know?*** (One record and two tapes cost less than two records and one tape. Both collections have one of each, so the different item determines the difference in price.)
- ***On the second sign, what could each item cost?*** (The CD could cost \$6 and the tape, \$5.) ***What other number pairs would work?*** (1, 10; 2, 9; 3, 8; 4, 7; 5, 6; 7, 4; 8, 3; 9, 2; and 10, 1)
- ***Could the tape cost \$5?*** (No) ***Why or why not?*** (If it cost \$5 on the first sign, then the record would cost \$4. On the third sign, if the record is \$4, then the tape would be \$8. That is a contradiction.)

Solutions

1 A record costs \$6.

2 A CD costs \$7.

3 A tape costs \$4.

4 One possible solution method:

Compare the items on the first and third signs. The cost of a record must be \$2 more than the price of a tape. One record and two tapes cost \$2 less than two records and one tape.
Since the difference in cost for the collections is \$2, the price of a record must be \$2 more than the price of a tape.
On the third sign, replace the tape with a third record. Since a record costs \$2 more than a tape, the price for these items increases by \$2. This makes three records cost \$18, and one record would cost \$6.
If a record costs \$6 and a tape costs \$2 less, a tape is \$4. On the second sign, if a tape costs \$4, then a CD costs \$7.

Variable

Tag Sale (E)

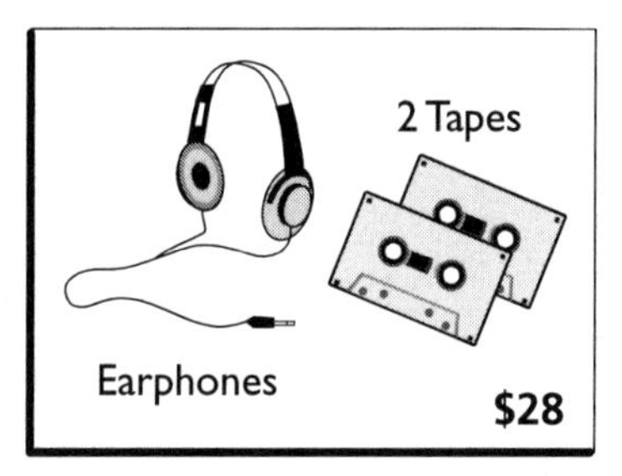

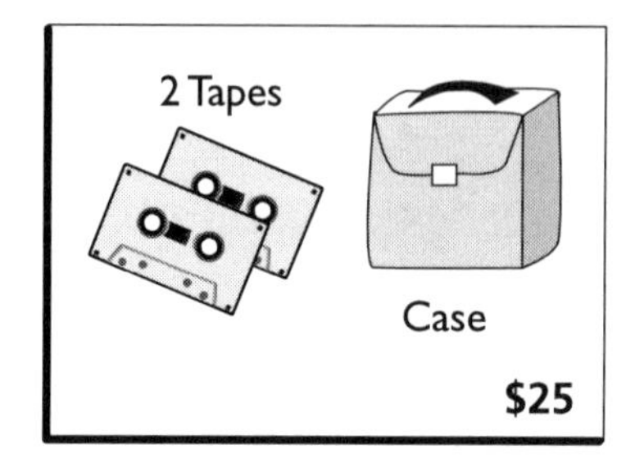

At a garage sale Shemetra is selling earphones, videotapes, and tape carrying cases. Same items have same prices. Different items have different prices. How much is:

1 One set of earphones? ______

2 One videotape? ______

3 One tape carrying case? ______

4 Explain how you figured out the prices.

Name ....

Permission is given by the publisher to the purchasing teacher or parent to reproduce this page for classroom or home use only.

Tag Sale (F)

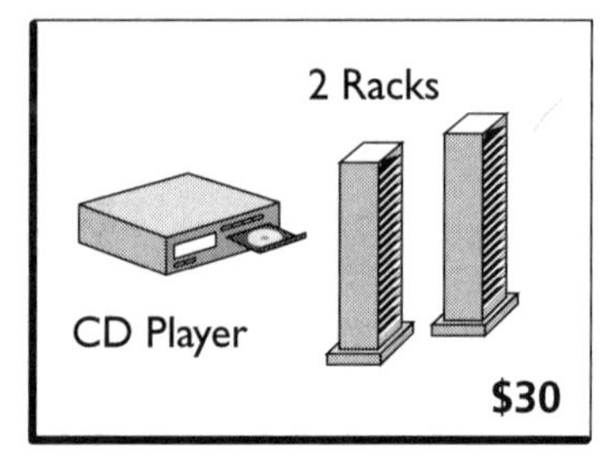

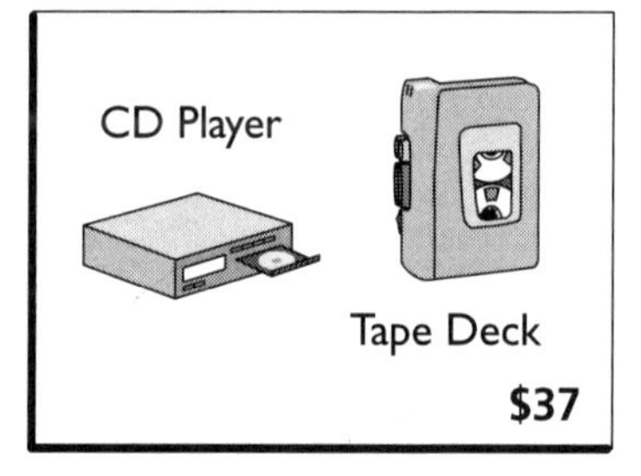

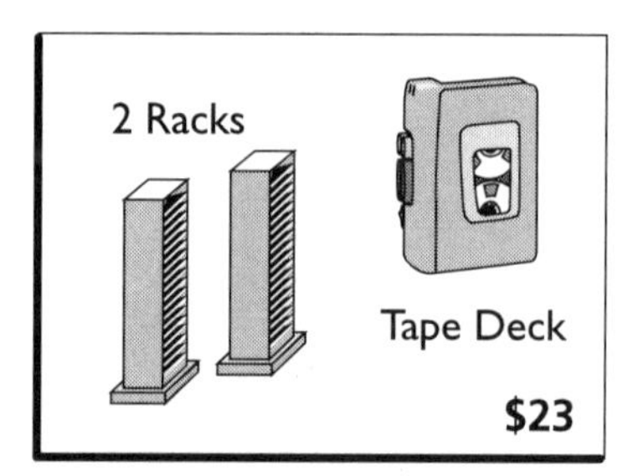

At a yard sale Kiesha is selling tape decks, CD players, and CD storage racks. Same items have same prices. Different items have different prices. How much is:

1 A tape deck? ______

2 A CD player? ______

3 A CD storage rack? ______

4 Explain how you figured out the prices.

Name ....

Permission is given by the publisher to the purchasing teacher or parent to reproduce this page for classroom or home use only.

© Creative Publications

Tag Sale (E)

Solutions

1 A set of earphones costs $8.

2 A videotape costs $10.

3 A tape carrying case costs $5.

4 One possible solution method:

On the second sign the earphones and a tape cost $18.
Replace the earphones and one tape with $18 on the first sign. The remaining tape is $10.
On the second sign, if a tape is $10, the earphones are $8.
On the third sign, if two tapes are $20, then the case is $5.

Tag Sale (F)

Solutions

1 A tape deck costs $15.

2 A CD player costs $22.

3 A storage rack costs $4.

4 One possible solution method:

Compare the items on the first and third signs. One CD player and two racks cost $7 more than a tape deck and two racks. Since the difference in cost for the collections is $7, the price of a CD player is $7 more than the price of a tape deck.

On the second sign, there is a CD player and a tape deck. Replace the tape deck with another CD player. Since the CD player costs $7 more than the tape deck, increase the price of these items by $7, to $44. So 1 CD player costs 44 ÷ 2, or $22.
If a CD player costs $22, and a tape deck costs $7 less, then a tape deck is $15.
On the third sign, if a tape deck costs $15, then the two racks cost 23 – 15, or $8, and one rack is $4.

© Creative Publications

Make a Table (A)

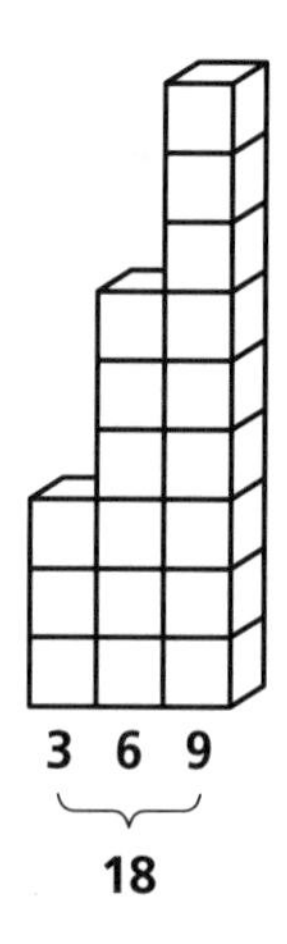

This is a staircase Deion built. He used 3 blocks for the first stair, 6 blocks for the second stair, 9 blocks for the third, and so on, using 3 more blocks for each higher stair.

1 Make a 3-column table. In Column 1 show the Stair Number. In Column 2 show Number of Blocks used. In Column 3 show the Total Number of Blocks used to build the stairs from Stair 1 up to and including Stair 5. For example, for Stairs 1–3, Deion used a total of 3 + 6 + 9, or 18, blocks.

2 Let *S* represent the Stair Number. Let *N* represent the Number of Blocks. Complete this function rule to show the number of blocks needed to build Stair *S*. $N =$ ______

3 What is the number of blocks Deion would need to build Stair 10? ______

4 Explain your answer to the above question.

Permission is given by the publisher to the purchasing teacher or parent to reproduce this page for classroom or home use only.

Name ....

© Creative Publications

Make a Table (A)

Goals

- Construct a three-column table to show related data.
- Continue a pattern.
- Describe in words a rule that compares pairs of data.
- Use symbols to generate a rule that describes a relationship.

Questions to Ask

- *How many blocks did Deion use to build the third stair?* (9)
- *How many blocks do you think he used for the fourth stair?* (12)
- *How many blocks in all did Deion need to build a 4-stair staircase?* (30)
- *How is the Stair Number related to the number of blocks needed to build it?* (The number of blocks is 3 times the Stair Number.)
- *Which stair would require 21 blocks to build?* (Stair 7) *How do you know?* (Each stair requires 3 times as many blocks as the Stair Number.)

Solutions

1

Stair	Blocks	Total Blocks
1	3	3
2	6	9
3	9	18
4	12	30
5	15	45

2 $N = 3S$

3 30 blocks

4 It's $3S$; in this case, 3×10, or 30.

© Creative Publications

Make a Table (B)

Sean read 60 pages on Monday, 57 pages on Tuesday, 54 on Wednesday, and so on, reading 3 fewer pages each day until he finished his book.

1 Make a 3-column table. In Column 1 show the Day Number. In Column 2 show the Number of Pages read daily. In Column 3 show the Total Number of Pages read from Day 1 up to and including Day 5.
For example, by the end of the third day, Sean had read 60 + 57 + 54, or 171, pages.

2 Let D represent the Day Number. Let P represent the Number of Pages read that day. Complete this function rule to show the number of pages read on Day D. $P =$ ______

3 What is the total number of pages Sean read in 6 days?

______

4 Sean finished his book the day he read 33 pages. How many pages were in the book? ______

Name ....

Permission is given by the publisher to the purchasing teacher or parent to reproduce this page for classroom or home use only.

Make a Table (C)

Carla sold 5 papers the first day of her job, 7 papers the second day, 9 papers the third, and so on, selling 2 more papers each day than on the day before.

1 Make a three-column table. In Column 1 show the Day Number. In Column 2 show the Number of Papers sold that day. In Column 3 show the Total Number of Papers sold from Day 1 up to and including Day 5. For example, by the end of Day 3, Carla had sold 5 + 7 + 9, or 21, papers.

2 Let D represent the Day Number. Let P represent the Number of Papers Sold. Complete this function rule to show the number of papers sold on Day D. $P =$ ______

3 What is the total number of papers Carla sold in 10 days?

______

4 Let D represent the Day Number. Let T represent the Total Number of Papers sold. Complete this function rule to show the number of papers sold, up to and including Day D.
$T =$ ______

Name ....

Permission is given by the publisher to the purchasing teacher or parent to reproduce this page for classroom or home use only.

© Creative Publications

Make a Table (B)

Solutions

1

Day	Pages	Total Pages
1	60	60
2	57	117
3	54	171
4	51	222
5	48	270

2 $P = 60 - 3(D - 1)$, or $P = 63 - 3D$

Sean read:

$60 - (3 \times 0)$ pages on Day 1, or 60

$60 - (3 \times 1)$ pages on Day 2, or 57

$60 - (3 \times 2)$ pages on Day 3, or 54

So Sean read

$60 - 3(D - 1)$ pages on Day D.

3 315 pages

4 465 pages - Compute the total number of pages read down through 33 pages. $60 + 57 + 54 + 51 + ... + 33 = 465$ pages.

Make a Table (C)

Solutions

1

Day	Papers	Total Papers
1	5	5
2	7	12
3	9	21
4	11	32
5	13	45

2 $P = 2D + 3$

3 140 papers

Two possible solution methods:

Compute the number of papers sold for each of the 10 days and then find the total. $5 + 7 + 9 + 11 + ... + 23 = 140$.

Look at the numbers in the Total Number of Papers column. Factor the numbers and look for patterns. Factoring shows that the numbers are: 1×5, 2×6, 3×7, 4×8, 5×9, etc. In each case, the first factor is the Day Number and the second factor is 4 more than the Day Number.

Write a rule to show how the factors of the Total relate to the Day Number: $T = D(D + 4)$. Use the rule to find the Total for 10 days: $T = 10(10 + 4)$, or 140, papers.

4 $T = D(D + 4)$

© Creative Publications

Make a Table (D)

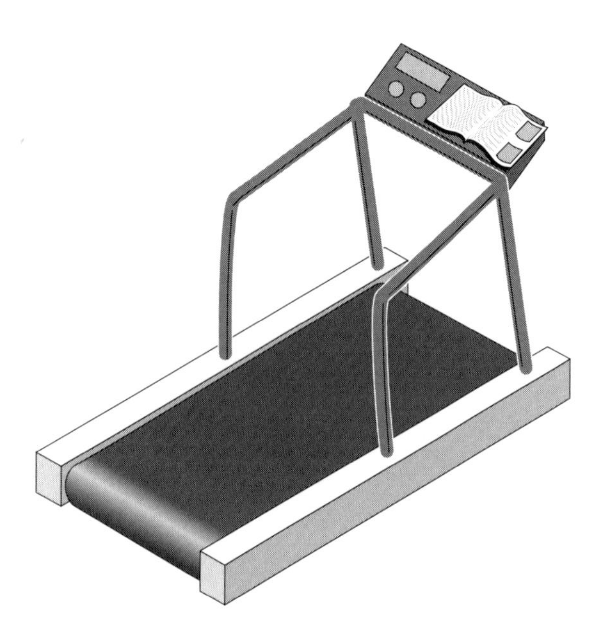

For aerobic exercise Jared walked on a treadmill 2 kilometers on Day 1, 4 kilometers on Day 2, 6 kilometers on Day 3, and so on, walking 2 more kilometers each day than on the day before.

1 Make a three-column table. In Column 1 show the Day Number. In Column 2 show the Number of Kilometers that Jared walked each day for the 10 days. In Column 3 show the Total Number of Kilometers that Jared walked from Day 1 up to and including Day 10. For example, on Day 3 Jared had walked a total of 2 + 4 + 6, or 12, kilometers.

2 Let D represent the Day Number. Let N represent the Number of Kilometers walked. Complete this function rule to show the number of kilometers that Jared walked each day. $N =$ ______

3 Write a rule in words that relates the total number of kilometers to the day number.

4 Let T represent the Total Number of Kilometers. Let D represent the Day Number. Complete the function rule that relates the total number of kilometers to the day number. $T =$ ______

Name ....

Permission is given by the publisher to the purchasing teacher or parent to reproduce this page for classroom or home use only.

 © Creative Publications

Make a Table (D)

Goals

- Construct a three-column table to show related data.
- Continue a pattern.
- Write in words a rule that relates pairs of data.
- Use symbols to generalize a relationship.

Questions to Ask

- *How many kilometers did Jared walk on Day 4?* (8) *On Day 5?* (10)
- *What is an easy way to figure out the number of kilometers Jared walked each day?* (Multiply the Day Number by 2.)
- *Beginning with Day 1 how many days will it take Jared to walk 30 kilometers?* (5) *How do you know?* (2 + 4 + 6 + 8 + 10 = 30)

Solutions

1

Day	Km/Day	Total Km
1	2	2
2	4	6
3	6	12
4	8	20
5	10	30
6	12	42
7	14	56
8	16	72
9	18	90
10	20	110

2 $N = 2D$

3 The total number of kilometers walked is equal to the day number times one more than the day number.

T = Day Number x (Day Number + 1), or $T = D^2 + D$

4 $T = D \times (D + 1)$, or $T = D^2 + D$

Make a Table (E)

Juanita did 1 sit-up on Day 1, 3 sit-ups on Day 2, 5 sit-ups on Day 3, and so on, doing 2 more sit-ups each day than on the previous day.

1 Make a three-column table. In Column 1 show the Day Number for 1 through 10 days. In Column 2 show the Number of Sit-Ups that Juanita did each day. In Column 3 show the Total Number of Sit-Ups from Day 1 up through Day 10.

2 Let D represent the Day Number. Let N represent the Number of Sit-Ups in a day. Let T represent the Total Number of Sit-Ups from Day 1 through that day. Write a function rule that relates N and D. $N =$ ______

3 Write a function rule that relates T and D. $T =$ ______

4 What was the total number of sit-ups that Juanita did in the first 20 days? ______

Name ....

Permission is given by the publisher to the purchasing teacher or parent to reproduce this page for classroom or home use only.

Make a Table (F)

Brandon jogged 1 block on Day 1 of his exercise program, 2 blocks on Day 2, 3 blocks on Day 3, and so on, jogging 1 more block each day than on the previous day.

1 Make a three-column table. In the first column, record the Day Number for 1 through 10 days.
In the second column, record the Number of Blocks that Brandon jogged each day for 10 days. In the third column, record the Total Number of Blocks that Brandon jogged from Day 1 up through Day 10.

2 Let D represent the Day Number. Let N represent the Number of Blocks jogged on any day. Let T represent the Total Number of Blocks that Brandon jogged beginning with Day 1. Write a function rule relating N and D. $N =$ ______

3 How does D relate to T? Explain how you can find one if you know the other.

4 What was the total number of blocks that Brandon jogged in the first 15 days? ______

Name ....

Permission is given by the publisher to the purchasing teacher or parent to reproduce this page for classroom or home use only.

© Creative Publications

Make a Table (E)

Solutions

1

Day	Sit-Ups	Total Sit-Ups
1	1	1
2	3	4
3	5	9
4	7	16
5	9	25
6	11	36
7	13	49
8	15	64
9	17	81
10	19	100

2 $N = 2D - 1$

3 $T = D \times D$ or D^2

4 If $D = 20$, then $T = 20 \times 20 = 400$. In 20 days Juanita did a total of 400 sit-ups.

Function

Make a Table (F)

Solutions

1

Day	Blocks	Total Blocks
1	1	1
2	2	3
3	3	6
4	4	10
5	5	15
6	6	21
7	7	28
8	8	36
9	9	45
10	10	55

2 $N = D$

3 Take the total number of blocks jogged for any day and add the next day's number. The result gives the total number of blocks jogged for the Day Number added. For example, for Day 3 the total number of blocks jogged is 6. Add 4, the next day's number. That gives 10, the total number of blocks jogged for Day 4. An equation that gives the same result is: $T = \frac{D(D+1)}{2} = \frac{D^2 + D}{2}$.

4 In 15 days Brandon jogged 120 blocks.

Function

© Creative Publications

Beginning and End (A)

Last Will and Testament of
Great Grandpa McLurdy:

- *To my dear son Purdy, I leave half of my estate.*
- *To my dear niece Gertie, I leave half of what is left.*
- *To my cook McMurdy, I leave the remainder of my estate, $100,000.*

1 What was the value of Great Grandpa McLurdy's entire estate?

______

2 Explain how you figured out the estate's value.

Name ....

Permission is given by the publisher to the purchasing teacher or parent to reproduce this page for classroom or home use only.

© Creative Publications

Beginning and End (A)

Goals

◆ Use inverse operations to compute inputs when given outputs.

Questions to Ask

◆ *After giving money to Purdy, what part of the total estate is left?* (One half)

◆ *After giving money to Gertie, what part of the total estate is left?* (Half of the half, or one fourth)

◆ *Suppose there was $20,000 in McLurdy's estate. How much money would Purdy get?* (Half of $20,000, or $10,000)

◆ *How much money would be left?* ($10,000)

◆ *How much would Gertie get?* (Half of $10,000, or $5000)

Solutions

1 The value of Great Grandpa McLurdy's estate was $400,000.

2 One possible solution method:

Start at the end and work backwards. Use inverse operations. If Gertie got half of what was left, then the other half went to McMurdy. Its value was $100,000, so Gertie also received $100,000. The value of what these two received was 2 × $100,000, or $200,000. This was what was left after Purdy got his share, so Purdy got $200,000. The value of the estate before Purdy received half was 2 × $200,000, or $400,000.

Function

Beginning and End (B)

Last Will and Testament of
Loris Lorinda:

- *To my sister Cloris, I leave one half of my estate.*
- *To my brother Horace, I leave one half of what is left.*
- *To my cousin Boris, I leave one half of what is left.*
- *To the care of my dog, Torus, I leave the remaining $20,000.*

1 What was the value of Loris Lorinda's estate? ______

2 Explain how you figured out the estate's value.

Name ....

Permission is given by the publisher to the purchasing teacher or parent to reproduce this page for classroom or home use only.

Beginning and End (C)

Last Will and Testament of
Brunhilda Brumkin:

- *To my Aunt Matilda, I leave one-third of my estate.*
- *To my cousin Gruzilda, I leave half of what is left.*
- *For the care of my parrot Palumpkin, I leave the remaining $40,000.*

1 What was the value of Brunhilda Brumkin's estate? ______

2 Explain how you figured out the estate's value.

Name ....

Permission is given by the publisher to the purchasing teacher or parent to reproduce this page for classroom or home use only.

© Creative Publications

Beginning and End (B)

Solutions

1 The value of Loris Lorinda's estate was $160,000.

2 One possible solution method:

If Torus got $20,000, which was one half of the remaining money, then Boris also received $20,000. Before Boris got his share, the value of the estate was 2 × $20,000, or $40,000. If Boris got half of what was left after Horace took one half, Boris got half of $40,000. Horace received $40,000. Before Horace took his share, the amount of money left after Cloris took one half was $80,000. That means that half of the estate's value was $80,000, and the whole estate was worth 2 × $80,000, or $160,000.

Function

Beginning and End (C)

Solutions

1 The value of Brunhilda Brumkin's estate was $120,000.

2 Two possible solution methods:

If Gruzilda got half of what was left, then Palumpkin got the other half. Since Palumpkin got $40,000, Gruzilda also received $40,000. Before Gruzilda got her share, there was $80,000 which was left after Aunt Matilda got her share. Since Matilda got one third, after she took her share two thirds of the estate was left. Two thirds of the estate was $80,000. If two thirds of the estate was $80,000, then the whole estate was $\frac{3}{2}$ × $80,000, or $120,000.

Each person actually received one third of the total estate. $40,000 is one third of what number? The total estate was worth $120,000.

Function

© Creative Publications

Beginning and End (D)

Dear Grandmother,

Thank you for the birthday money. First, I spent half the money for clothes. Then I spent $6 to go to a movie. Then I spent half of what was left for a CD. Now I have $7 left.

Love,
Pietra

1 How much money did Grandmother give Pietra? ______

2 Explain how you figured out the answer.

Name ....

Permission is given by the publisher to the purchasing teacher or parent to reproduce this page for classroom or home use only.

© Creative Publications

Beginning and End (D)

Goals

- Use inverse operations to get the input when given the output.

Questions to Ask

- ***How much money did Pietra have left when she wrote the thank-you note?*** ($7)
- ***How much money did Pietra spend for the CD?*** ($7)
- ***How do you know?*** (The CD cost half of her remaining money, and she was left with $7 after she bought it.)
- ***How much of the total amount did Pietra spend for clothes?*** (Half)
- ***Where would be a good place to start solving this problem?*** (Begin at the end, and work backwards from the $7 Pietra had left.)

Solutions

1 Grandmother gave Pietra $40.

2 One possible solution method:

Pietra had $7 left. The last thing she did was spend half her remaining money for a CD. Since she had $7 left, the CD must have cost $7. She had $14 before she bought the CD. Before she bought the CD, Pietra spent $6 for a movie. She had 14 + 6, or $20, before the movie. After buying clothes, Pietra had $20. She spent half of her money on clothes, so they must have cost $20, too. Grandmother gave Pietra 20 + 20, or $40.

© Creative Publications

Beginning and End (E)

Name ....

1 How much money did Uncle Gus give Jan? ______

2 Explain how you figured out the answer.

Dear Uncle Gus,

Thank you for the money gift. First I spent one third of the money for a concert ticket. Then I spent half of what was left to buy a gift for my best friend. Next I spent $6 for a CD. Last I spent half of what was left for bus fare to the concert. Now I have $5 left.

Love,
Jan

Permission is given by the publisher to the purchasing teacher or parent to reproduce this page for classroom or home use only.

Beginning and End (F)

Name ....

1 How much money did Aunt Chloe give Alfie? ______

2 Explain how you figured out the answer.

Dear Aunt Chloe,

Thank you for the money gift. First I spent half the money for books. Then I spent $3 for a new notebook. I spent half of what was left for lunch at a restaurant. Then I spent half of what was left for a card for Mom's birthday. Now I have $3 left.

Love,
Alfie

Permission is given by the publisher to the purchasing teacher or parent to reproduce this page for classroom or home use only.

© Creative Publications

Beginning and End (E)

Solutions

1 Uncle Gus gave Jan $48.

2 One possible solution method:

Jan had $5 left. The last thing she did was spend half her remaining money for bus fare. Since she was left with $5, the bus fare must have cost $5. She had $10 before that. Before buying the ticket Jan spent $6 for tapes. She had 10 + 6, or $16, before the purchase. After buying the gift for her friend, Jan had $16. She spent half her money for the gift, so it must have cost $16. Jan had $32, before she bought the gift.

After Jan spent one third of her money for the concert ticket, she had $32. If the ticket cost one third of the gift money, the remaining $32 must be the other two thirds of the gift money. If two thirds is $32, then one third must be $16, and the whole gift must have been 3 × 16, or $48.

Function

Beginning and End (F)

Solutions

1 Aunt Chloe gave Alfie $30.

2 One possible solution method:

Alfie had $3 left. The last thing he did was spend half his remaining money for a birthday card. Since he was left with $3, the card must have cost $3. Alfie had $6 before he bought the card. Before buying the card Alfie spent half his money for lunch. Since he had $6 after lunch, lunch must have cost $6. Before lunch, Alfie had $12.

After buying the notebook Alfie had $12. He spent $3 for it, so he had 12 + 3, or $15, before buying the notebook. After Alfie bought books, he had $15. He spent half of his gift money for books, so they must have cost $15. Alfie had 15 + 15, or $30, before he bought the books. Aunt Chloe gave him $30.

Function

Functions and Graphs (A)

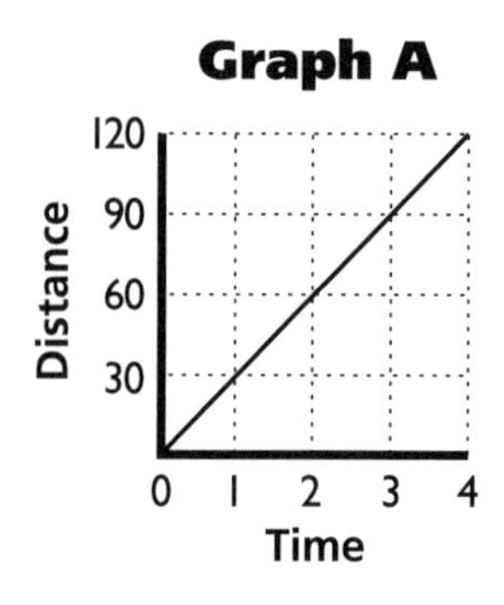

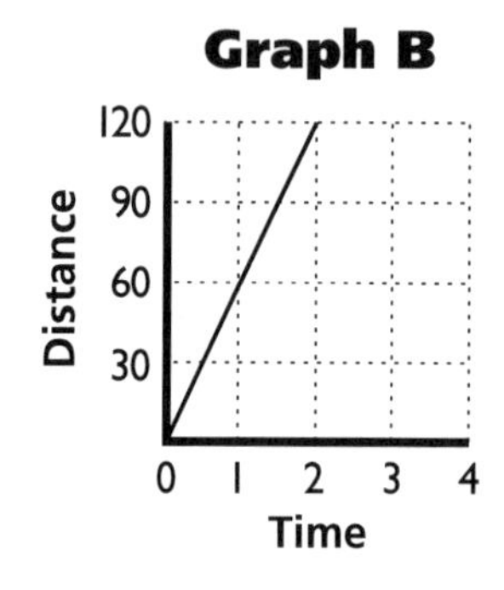

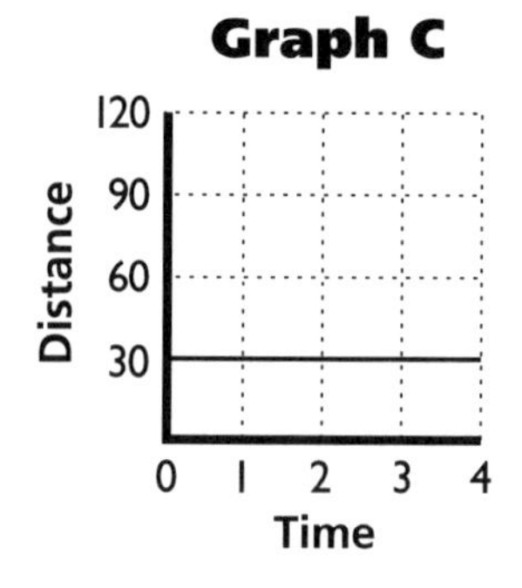

These graphs represent three different trips.
Nate left home and drove at an average speed of 30 mph.
Let T represent the number of hours.
Let D represent the number of miles traveled.

1 Which graph, A, B, or C, shows this distance-time relationship?

______

2 Write a function rule that relates T to D. $D =$ ______

3 Use the function rule. How many miles did Nate drive in 6 hours?

______

Name ....

Permission is given by the publisher to the purchasing teacher or parent to reproduce this page for classroom or home use only.

© Creative Publications

Functions and Graphs (A)

Function

Goals

- Match a linear time-distance graph to a mathematical relationship presented in words.
- Use symbols to write a rule that relates time to distance.
- Use an equation to solve a problem.

Questions to Ask

- *What does each graph show?* (The distance traveled and the time it took to do so)
- *Why do the graphs look different?* (Each one shows a different distance-time relationship.)
- *In Graph A how many miles did the traveler go in 2 hours?* (60 miles)
- *In Graph B how many hours did it take someone to travel 90 miles?* (1.5 hours)
- *In Graph C how many miles did a person travel in 4 hours?* (30 miles)

Solutions

1. Graph A - It shows that the distance is 30 miles in 1 hour, 60 miles in 2 hours, and 90 miles in 3 hours. The average is 30 miles per hour.
2. $D = 30T$
3. If $T = 6$, then $D = 30 \times 6$, or 180 miles.

© Creative Publications

Functions and Graphs (B)

Name

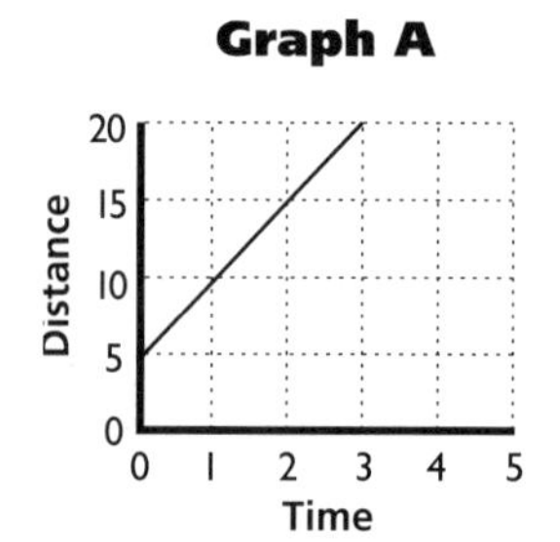

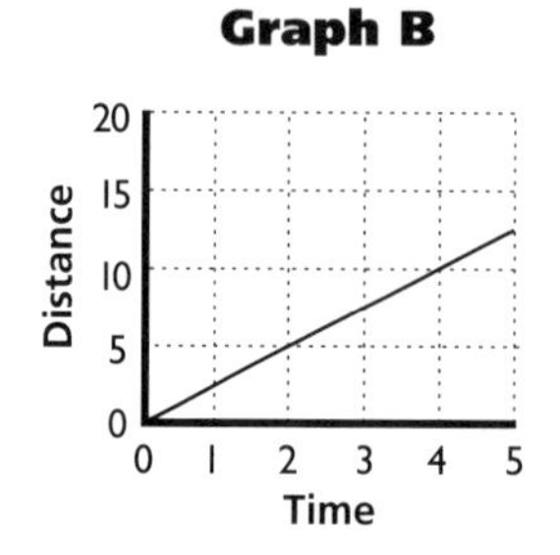

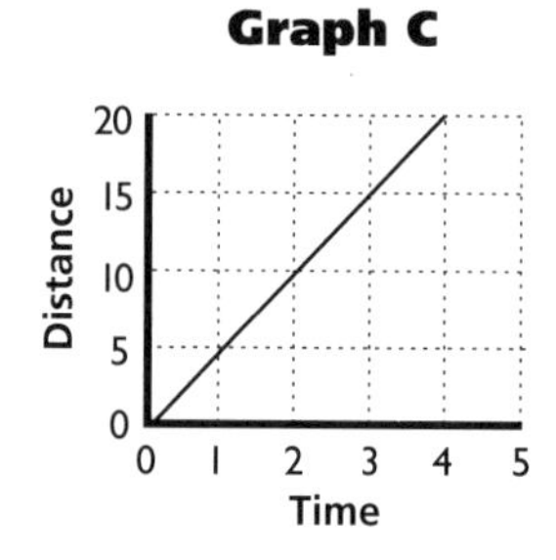

These graphs represent three different bike trips. Julian left home and biked at an average speed of 5 miles per hour.

1 Which graph, A, B, or C, shows this distance-time relationship? ______

2 Write a function rule to relate *D*, the total distance traveled, to *T*, the number of hours Julian biked. *D* = ______

3 Use the function rule. How many miles did Julian bike in 5 hours? ______

Permission is given by the publisher to the purchasing teacher or parent to reproduce this page for classroom or home use only.

Functions and Graphs (C)

Name

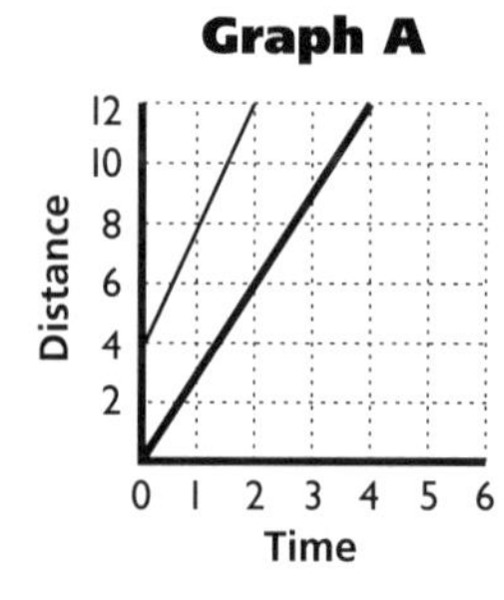

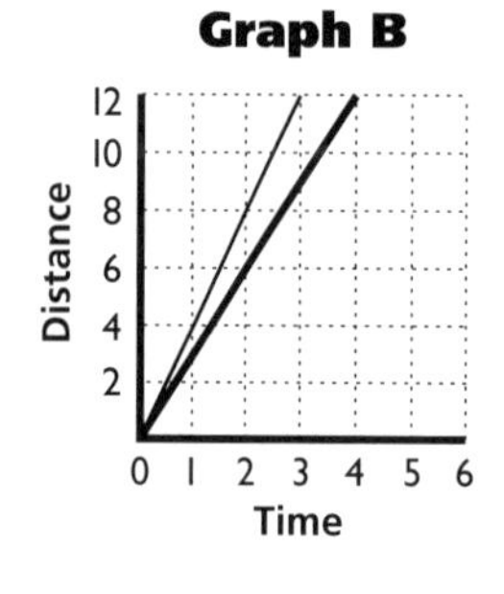

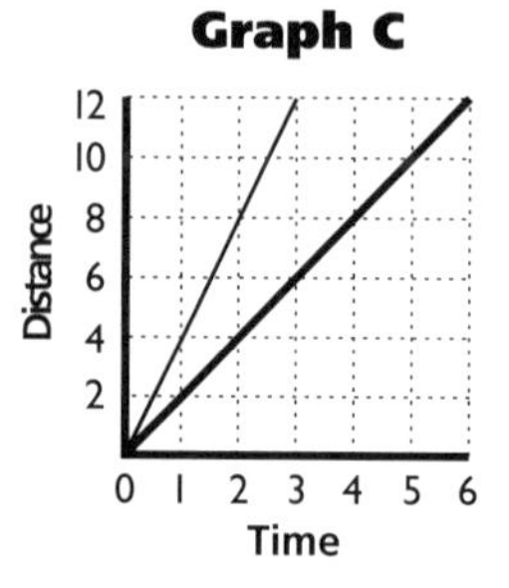

These graphs represent three different jogging periods. Allison and Danya are sisters. Allison left home and jogged at a speed of 3 miles per hour. Danya left home at the same time and jogged at a speed of 4 miles per hour.

1 Which graph, A, B, or C, represents Allison and Danya's trips? ______

2 Write a function rule to relate *T*, the number of hours Allison jogged, to *D*, the distance she traveled. *D* = ______

3 Write a function rule to relate *T*, the number of hours Danya jogged, to *D*, the distance she traveled. *D* = ______

Permission is given by the publisher to the purchasing teacher or parent to reproduce this page for classroom or home use only.

© Creative Publications

Functions and Graphs (B)

Solutions

1 Graph C shows an average speed of 5 mph.

2 $D = 5T$

3 If $T = 5$, then $D = 5 \times 5$, or 25 miles.

Function

Functions and Graphs (C)

Solutions

1 Graph B - It shows that Allison jogged 3 miles per hour, and Danya jogged 4 miles per hour. Since both girls left at the same time from the same place, both lines have to begin at the origin, (0, 0). Only Graph B shows that at Hour 1, one jogger is at 3 miles and the other jogger is at 4 miles.

2 $D = 3T$ for Allison

3 $D = 4T$ for Danya

Function

Functions and Graphs (D)

Graph A

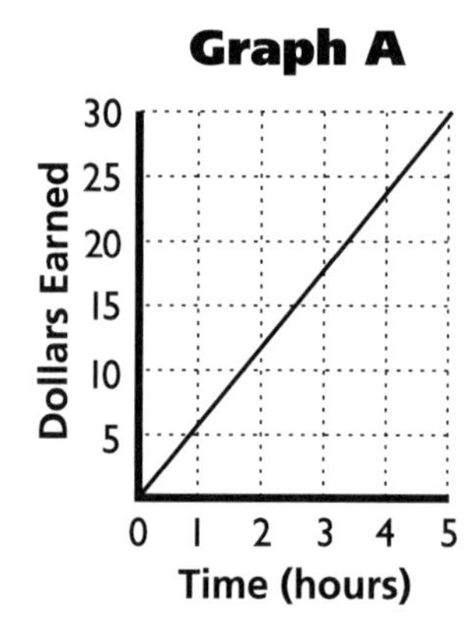

Graph B

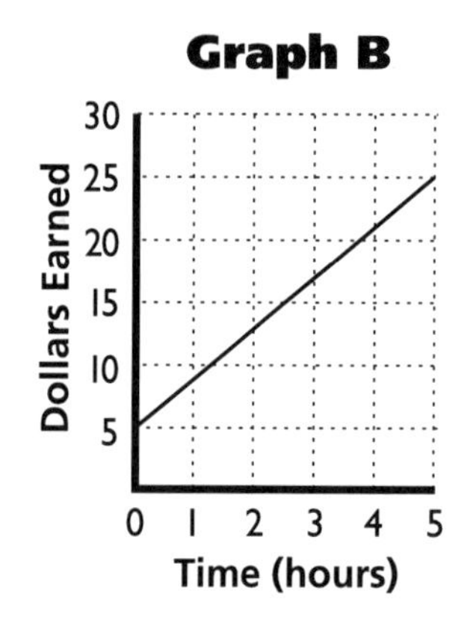

Graph C

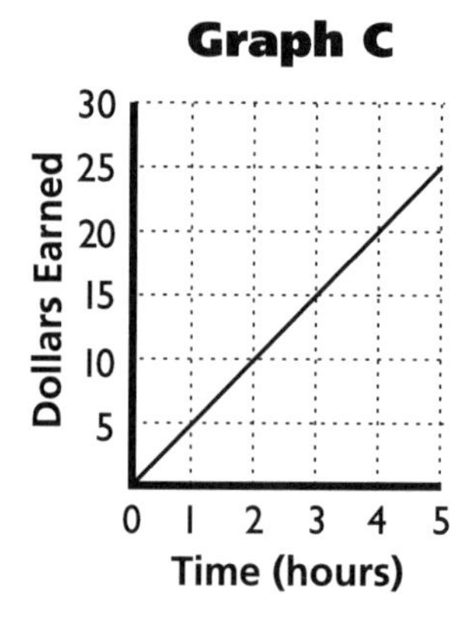

Samantha works as a helper at a daycare center. She earns $5 an hour.

1 Which graph, A, B, or C, shows this time-money relationship?

______

2 Write a function rule that relates *T* and *D*.
Let *T* represent the number of hours worked. Let *D* represent the number of dollars earned. $D =$ ______

3 Use the function rule. How much money did Samantha earn for 8 hours of work? ______

4 Explain how you chose the right graph.

Name ....

Permission is given by the publisher to the purchasing teacher or parent to reproduce this page for classroom or home use only.

© Creative Publications

Functions and Graphs (D)

Goals

- Use symbols to write a rule that relates time to money earned.
- Match a linear time-money graph to a mathematical relationship presented in words.
- Use an equation to solve a problem.

Questions to Ask

- *How much does Samantha earn in 2 hours?* ($10)
- *How many hours would Samantha have to work to earn $15?* (3)
- *In Graph A how much money does someone earn for 5 hours of work?* ($30)
- *What seems odd about Graph B?* (It shows someone earning $5 for no hours of work.)

Solutions

1 Graph C

2 $D = 5T$

3 Samantha earned 5×8, or $40, in 8 hours.

4 Graphs A and C appear to show about $5 earned for 1 hour of work, but Graph A shows $30 earned for 5 hours of work, which averages to $6 per hour. Samantha earned $5 per hour, so Graph A cannot be correct. Graph C shows $5 for 1 hour and $25 for 5 hours.

© Creative Publications

Functions and Graphs (E)

Name ....

Graph A

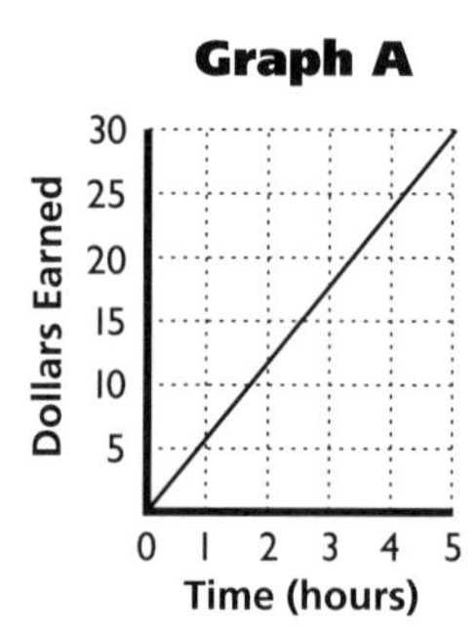

Graph B

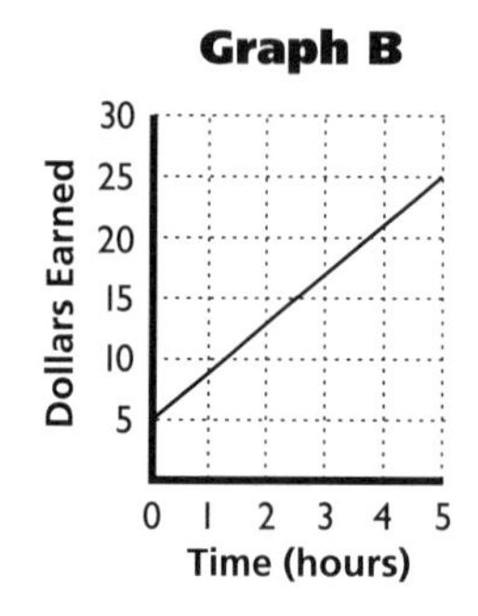

Graph C

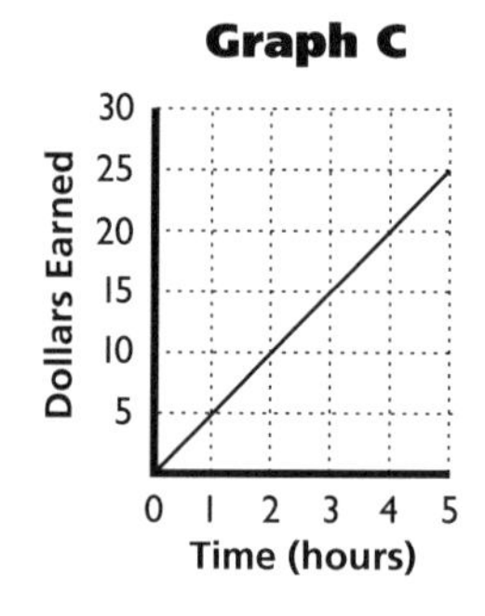

Joel works at a fast food restaurant. He earns \$6 an hour.

1 Which graph, A, B, or C, shows this time-money relationship? ______

2 Write a function rule that relates *T* and *D*. Let *T* represent the number of hours worked. Let *D* represent the number of dollars earned. *D* = ______

3 Use the function rule. How much money did Joel earn for 10 hours of work? ______

Permission is given by the publisher to the purchasing teacher or parent to reproduce this page for classroom or home use only.

Functions and Graphs (F)

Name ....

Graph A

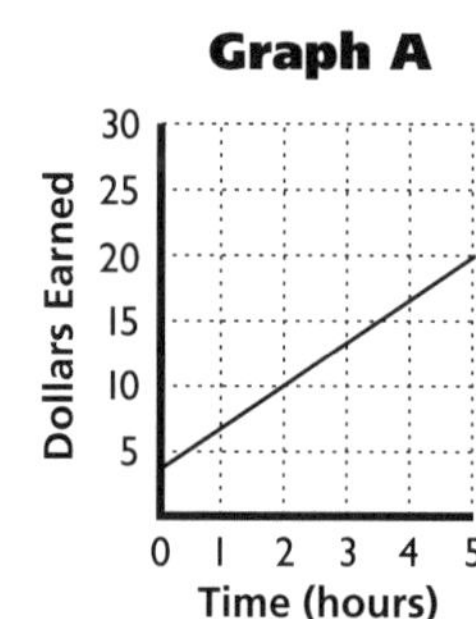

Graph B

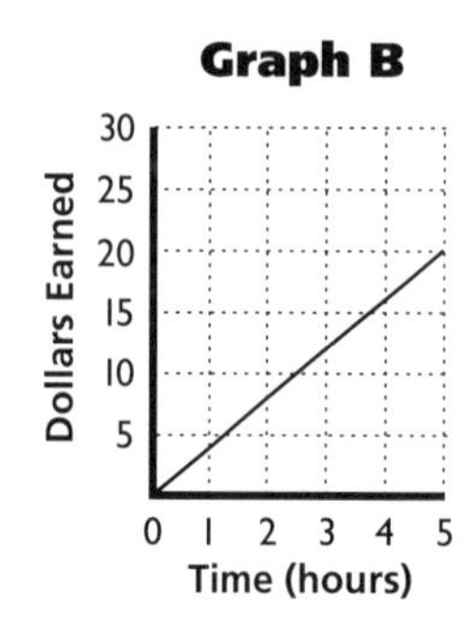

Graph C

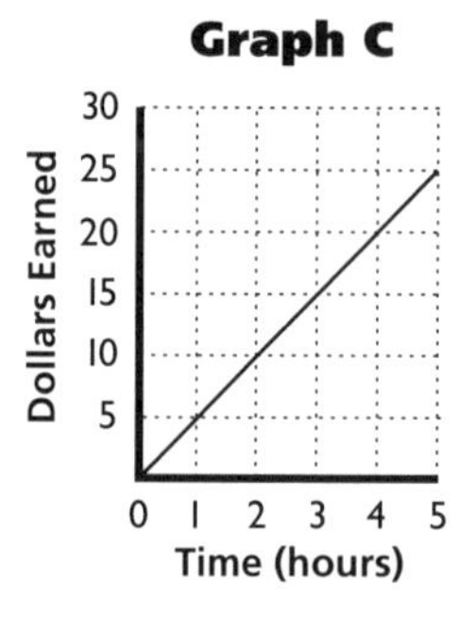

Andrea mows lawns for her neighbors. She earns \$4 an hour.

1 Which graph, A, B, or C, shows this time-money relationship? ______

2 Write a function rule that relates *T* and *D*. Let *T* represent the number of hours worked. Let *D* represent the number of dollars earned. *D* = ______

3 Use the function rule. How much does Andrea earn for 7 hours of work? ______

Permission is given by the publisher to the purchasing teacher or parent to reproduce this page for classroom or home use only.

© Creative Publications

Functions and Graphs (E)

Solutions

1 Graph A - Graph B shows \$5 earned for 0 hours of work, so Graph B is not correct. Graph C shows \$5 earned for 1 hour of work and \$25 earned for 5 hours of work, but Joel earned \$6 each hour. Check Graph A. It shows \$6 for 1 hour, \$30 for 5 hours, or \$6 per hour.

2 $D = 6T$

3 In 10 hours Joel earned 6×10, or \$60.

Functions and Graphs (F)

Solutions

1 Graph B - Graph A shows about \$6 earned for 1 hour of work, so Graph A is not correct. Graph C shows \$5 earned for 1 hour of work and \$25 earned for 5 hours of work, but Andrea earned only \$4 each hour. Check Graph B. It shows \$4 for 1 hour and \$20 for 5 hours of work.

2 $D = 4T$

3 In 7 hours Andrea earned 4×7, or \$28.

Keep Building (A)

Shape **1** **2** **3** **4**

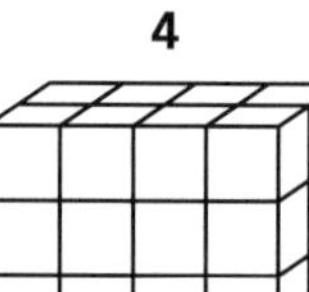

2 blocks 8 blocks 18 blocks 32 blocks

This is a series of growing shapes. Imagine that the building pattern continues. How many blocks will it take to build:

1 Shape 6? ______

2 Shape 10? ______

3 Tell in words how you would build Shape 5.

4 Write a function rule to relate the number of blocks in a shape to the shape number. Let *N* represent the shape number. Let *B* represent the number of blocks in a shape. $B =$ ______

Name ....

Permission is given by the publisher to the purchasing teacher or parent to reproduce this page for classroom or home use only.

© Creative Publications

Keep Building (A)

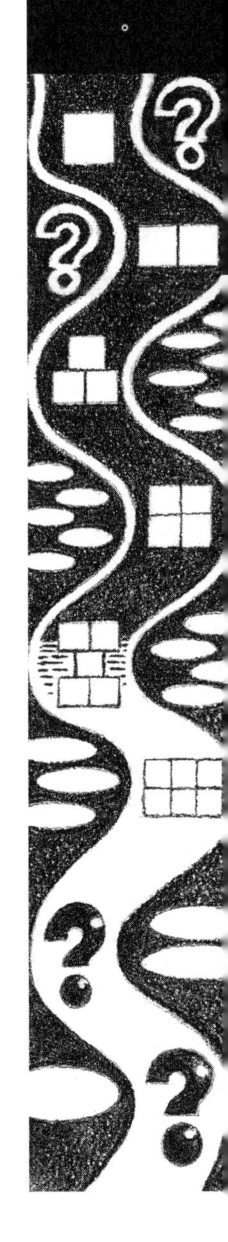

Goals

- Identify and continue patterns.
- Generalize patterns and relationships using words.
- Use symbols to create rules that relate pairs of data.

Questions to Ask

- *How many layers of blocks are in Shape 1?* (One) *How many blocks are in one layer?* (Two)
- *How many blocks are in Shape 1 in all?* (Two)
- *How many layers of blocks are in Shape 2?* (Two) *How many blocks are in one layer?* (Four) *How many blocks in Shape 2 in all?* (Eight)
- *How many layers of blocks are in Shape 3?* (Three) *How many blocks are in one layer?* (Six) *How many blocks in Shape 3 in all?* (18)
- *How are the shapes alike?* (They all are two rows deep; the number of layers is equal to the shape number; the number of blocks in a layer is twice the shape number.)
- *How are the shapes different?* (They are different sizes; they have different numbers of layers and different numbers of blocks in a layer; the number of blocks to build each shape differs.)

Solutions

1. There are 6 layers of 12 blocks, 6×12, or 72, blocks in Shape 6.
2. There are 10 layers of 20 blocks, or 200, blocks in Shape 10.
3. Shape 5 will have 5 layers. Each layer will have 10 blocks in a 2-by-5 arrangement.
4. $B = N \times (N + N)$ or $N^2 + N^2$

© Creative Publications

Keep Building (B)

Name

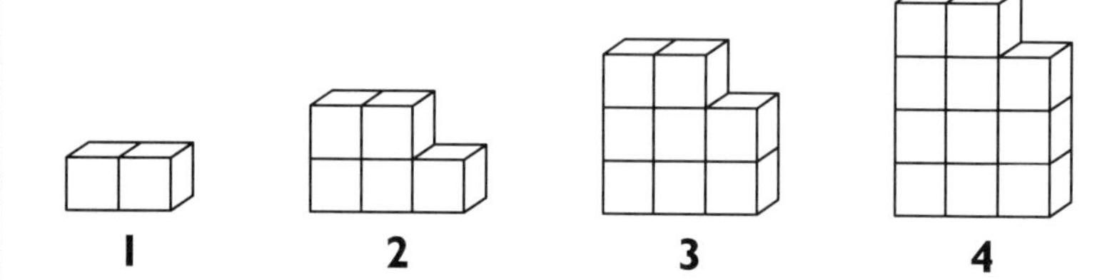

This is a series of growing shapes. Imagine that the building pattern continues. How many blocks will it take to build:

1 Shape 8? ______

2 Shape 12? ______

3 Tell in words how you would build Shape 5.

4 Write a function rule to relate the number of blocks in a shape to the shape number. Let N represent the shape number. Let B represent the number of blocks in a shape.
$B =$ ______

Permission is given by the publisher to the purchasing teacher or parent to reproduce this page for classroom or home use only.

Keep Building (C)

Name

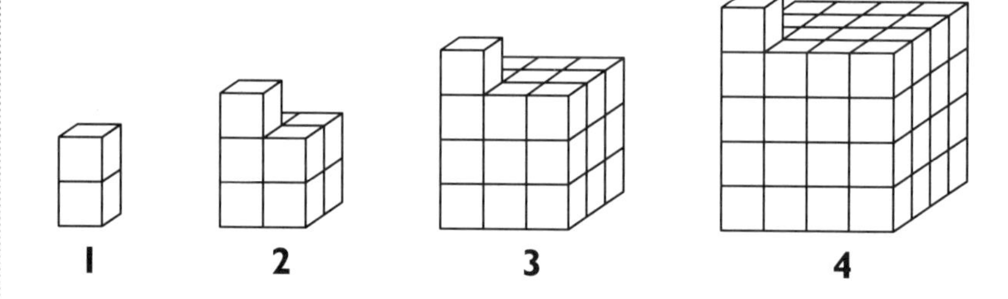

This is a series of growing cubes, with one small cube on top. Imagine that the building pattern continues. How many blocks will it take to build:

1 Cube 7? ______

2 Cube 20? ______

3 Tell in words how you would build Cube 5.

4 Write a function rule to relate the number of blocks in a cube to the cube number. Let N represent the cube number. Let B represent the number of blocks in a cube.
$B =$ ______

Permission is given by the publisher to the purchasing teacher or parent to reproduce this page for classroom or home use only.

© Creative Publications

Keep Building (B)

Solutions

1 There are 23 blocks in Shape 8.

2 There are 35 blocks in Shape 12.

3 There are 4 rows of 3 blocks in Shape 5. There are 2 blocks on the top, a total of 14 blocks.

4 $B = 2 + 3(N - 1) = 3N - 1$

Keep Building (C)

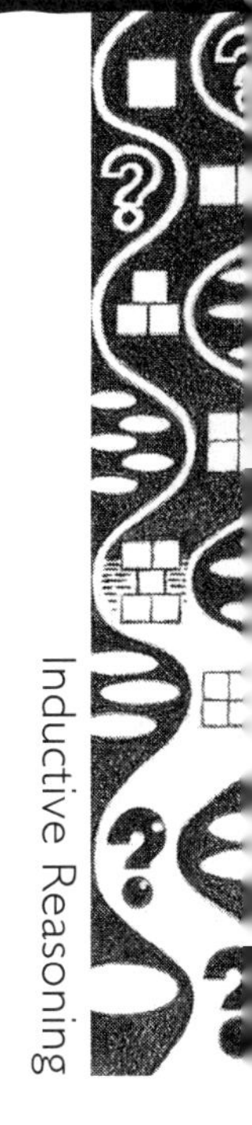

Solutions

1 There are $1 + 7^3$, or 344, blocks in Cube 7.

2 There are $1 + 20^3$, or 8001, blocks in Cube 20.

3 To build Cube 5 make a 5-by-5-by-5 cube with 125 blocks. Put 1 block on top.

4 $B = 1 + N^3$

© Creative Publications

Keep Building (D)

Shape **1** **2** **3** **4**

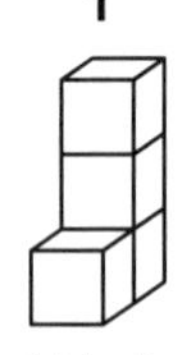
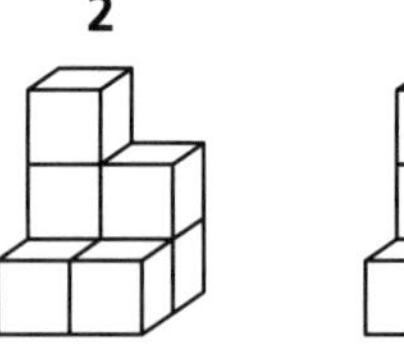
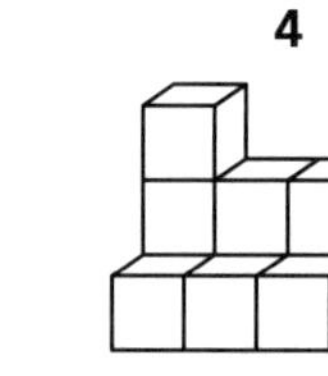

4 blocks 7 blocks 10 blocks 13 blocks

This is a series of growing shapes. Imagine that the building pattern continues. How many blocks will it take to build:

1 Shape 5? ______

2 Shape 10? ______

3 Write a function rule to find the number of blocks in a shape. Let *N* represent the shape number. Let *B* represent the number of blocks. *B* = ______

4 Which shape would take exactly 100 blocks to build? ______

Name ....

Permission is given by the publisher to the purchasing teacher or parent to reproduce this page for classroom or home use only.

© Creative Publications

Keep Building (D)

Goals

- ◆ Identify and continue patterns.
- ◆ Write a rule that relates a construction to its location in a sequence.
- ◆ Use symbols to create rules that relate pairs of data.

Questions to Ask

- ◆ *How many more blocks does each shape have than the shape right before it?* (Three)
- ◆ *How would you build Shape 5?* (Start with Shape 4 and put three blocks in the shape of an L next to the three blocks on the right side of the shape.)
- ◆ *What other way could you do it?* (Make five L-shapes with three blocks each. Then put one extra block on the top left corner.)
- ◆ *How are the shape number and the number of blocks related?* (The number of blocks is three times the shape number plus one.)

Solutions

1 There are $3 \times 5 + 1$, or 16, blocks in Shape 5.

2 There are $3 \times 10 + 1$, or 31, blocks in Shape 10.

3 The number of blocks in a shape is equal to one more than three times the Shape Number; $B = 3N + 1$.

4 Shape 33 would have $3 \times 33 + 1$, or exactly 100, blocks.

© Creative Publications

Keep Building (E)

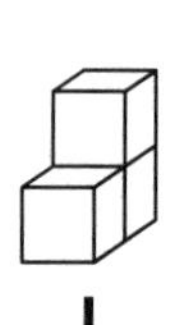
1

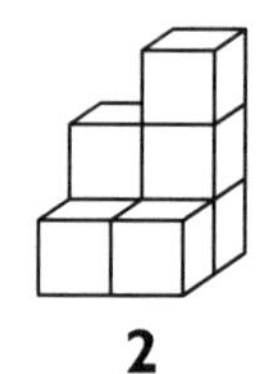
2

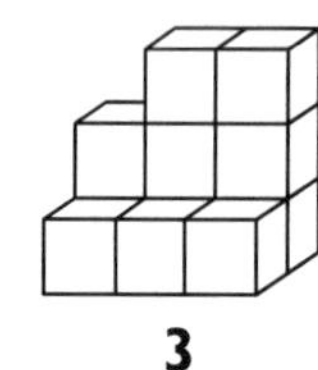
3

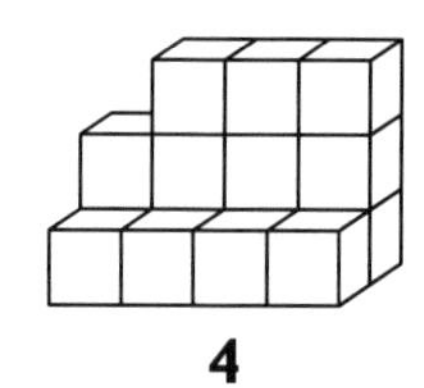
4

This is a series of growing shapes. Imagine that the building pattern continues. How many blocks will it take to build:

1 Shape 5? ______

2 Shape 8? ______

3 Write a function rule to find the number of blocks in a shape. Let N represent the shape number. Let B represent the number of blocks. $B =$ ______

4 Which shape would take exactly 107 blocks to build?

______

Name ....

Permission is given by the publisher to the purchasing teacher or parent to reproduce this page for classroom or home use only.

Keep Building (F)

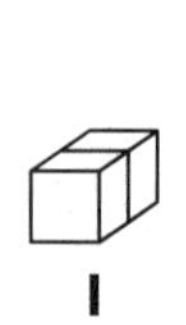
1

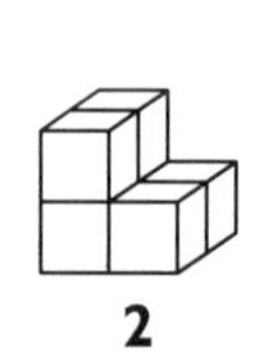
2

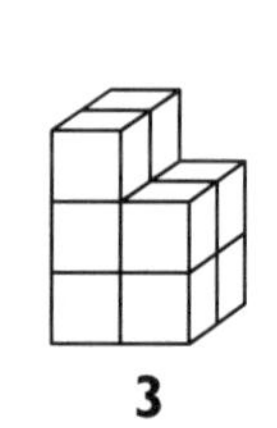
3

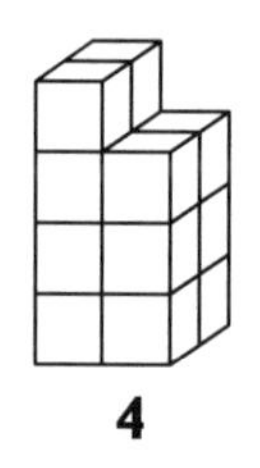
4

This is a series of growing shapes. Imagine that the building pattern continues.

1 How many blocks will it take to build Shape 10? ______

2 Write a function rule to find the number of blocks in a shape. Let N represent the shape number. Let B represent the number of blocks. $B =$ ______

3 How many blocks would be in all of the shapes, from Shape 1 up to and including Shape 10? ______

4 Write a function rule for finding the number of blocks in all of the shapes. Let N represent the shape number. Let T represent the total number of blocks. $T =$ ______

Name ....

Permission is given by the publisher to the purchasing teacher or parent to reproduce this page for classroom or home use only.

© Creative Publications

Where's the Seat? (A)

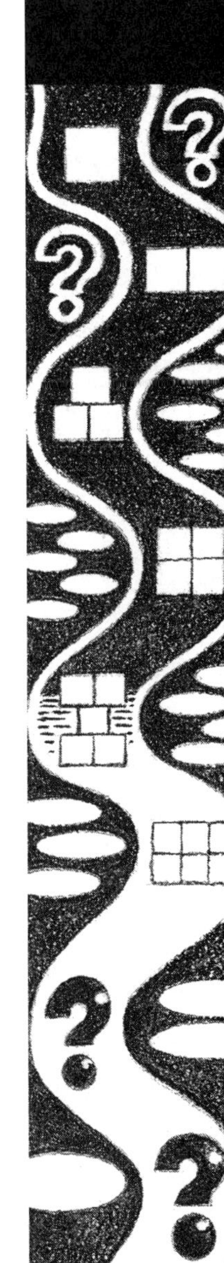

Goals

- Identify and continue patterns.
- Generalize patterns and relationships.
- Use symbols to create rules that relate pairs of data.

Questions to Ask

- *What is the number of the first seat in Row 4?* (19)
- *How many seats are in Row 1?* (4) *In Row 2?* (6) *In Row 3?* (8)
- *Do you see any pattern in the number of seats in the rows?* (Yes) *What is it?* (The numbers of seats are consecutive even numbers beginning with 4.)
- *How many seats will there be in Row 5?* (12)
- *How many seats do you think will be in Row 10?* (2 times the [Row Number + 1]. In this case: $2[R + 1] = 2 \times [10 + 1] = 22$ seats.)

Solutions

1 There are 18 seats in Row 8. You could count up by twos from Row 4 (10 seats) to Row 8.

2 $S = 2 \times (R + 1)$ or $2R + 2$

3 Seat 74 is in Row 8.

4 One possible solution method is to make a table:

Row	Last Number
1	$4 = 1 \times 4$
2	$10 = 2 \times 5$
3	$18 = 3 \times 6$
4	$28 = 4 \times 7$
5	$40 = 5 \times 8$
R	$L = R \times (R + 3)$

Test different values of R to find two values for L, one less than and one more than 74. Try $R = 7$, then $L = 7 \times (7 + 3) = 70$, and $L = 8 \times (8 + 3) = 88$. Since the last number in Row 7 is 70, then Seat 74 is in Row 8.

© Creative Publications

Where's the Seat? (B)

Rows							
4	10	11	12	13	14	15	16
3		5	6	7	8	9	
2			2	3	4		
1				1			

This diagram shows some of the seats in one section of a baseball stadium. The pattern of seats continues through Row 20.

1 How many seats are in Row 10? ______

2 Write a function rule to relate the number of seats in a row to the row number. Let R represent the row number. Let S represent the number of seats in a row.
$S =$ ______

3 Write a function rule to relate the last number in a row to the row number. Let R represent the row number.
Let L represent the last number in a row. $L =$ ______

4 Brenda is sitting in seat 121. In which row is her seat? ______

Name ....

Permission is given by the publisher to the purchasing teacher or parent to reproduce this page for classroom or home use only.

Where's the Seat? (C)

Rows								
4	13	14	15	16	17	18	19	20
3		7	8	9	10	11	12	
2			3	4	5	6		
1				1	2			

This diagram shows some of the seats in one section of a hockey rink. The pattern of seats continues through Row 20.

1 How many seats are in Row 20? ______

2 Write a function rule to relate S, the number of seats in a row, to R, the row number.
$S =$ ______

3 Write a function rule to relate L, the last number in a row, to R, the row number. $L =$ ______

4 Melinda is sitting in seat 100. In which row is her seat? ______

Name ....

Permission is given by the publisher to the purchasing teacher or parent to reproduce this page for classroom or home use only.

© Creative Publications

Where's the Seat? (B)

Solutions

1 There are 19 seats in Row 10.

2 The table shows the row numbers and the number of seats in each row.

Row Number	Number of Seats
1	1
2	3
3	5
4	7
5	9
6	11
R	$S = 2R - 1$

3 The table shows the row numbers and the number of the last seat in each row.

Row Number	Last Number
1	1
2	4
3	9
4	16
5	25
R	$L = R \times R$, or R^2

4 Since 121 is a square number and is the square of 11, Seat 121 is the last seat in Row 11.

Where's the Seat? (C)

Solutions

1 There are 40 seats in Row 20.

2 The table shows the row numbers and the number of seats in each row.

Row Number	Number of Seats
1	$2 = 1 \times 2$
2	$4 = 2 \times 2$
3	$6 = 3 \times 2$
4	$8 = 4 \times 2$
5	$10 = 5 \times 2$
R	$S = 2R$

3 The table shows the row numbers and the number of the last seat in each row.

Row Number	Last Number
1	$2 = 1 \times 2$
2	$6 = 2 \times 3$
3	$12 = 3 \times 4$
4	$20 = 4 \times 5$
5	$30 = 5 \times 6$
R	$L = R \times (R + 1)$ or $R^2 + R$

4 Try different values of R to find L values that are close to 100. The last number in Row 9 is 90. The last number in Row 10 is 110. Seat 100 is in Row 10.

© Creative Publications

Where's the Seat? (D)

This diagram shows some of the seats in one section of a concert hall. The pattern of seats continues through Row 25.

1 What is the number of the last seat in Row 10? ______
How do you know?

2 Write a function rule to relate the number of the last seat in a row to the row number. Let *R* represent the row number. Let *L* represent the number of the last seat in a row. $L =$ ______

3 Sean is sitting in Seat 70. In which row is his seat? ______

4 Describe how you found Sean's row number.

Name ....

Permission is given by the publisher to the purchasing teacher or parent to reproduce this page for classroom or home use only.

© Creative Publications

Where's the Seat? (D)

Goals

- Identify and continue patterns.
- Generalize patterns and relationships.
- Use symbols to create rules that relate pairs of data.

Questions to Ask

- ***How many numbers are in Row 5?*** (5)
- ***How many numbers are in the first five rows altogether?*** (15)
- ***What is the last number in Row 7?*** (28) ***How do you know?*** (The last number will be 7 more than the last number in Row 6, because the last numbers in the rows differ by 2, 3, 4, 5, 6, and then 7.)

Solutions

1 Seat 55 - The number at the end of each row tells the number of numbers in all rows up to and including that row. So, the end of the tenth row is found by computing the sum, $1 + 2 + 3 + \ldots + 10$. This sum is 55.

2 $L = 1 + 2 + 3 + \ldots + R$, or $L = [R(R + 1)] \div 2$

(L is the sum of all of the counting numbers from 1 up to and including the row number. This sum can be found by computing half of the product of the row number and one more than the row number.)

3 Sean is in Row 12.

4 Use the rule from above. Use guess and check to find the last number in two rows so that the number 70 is between these two numbers. The last number in Row 10 is 55. Sean has a row number greater than 10. Find the last number in Row 11. It is $(11 \times 12) \div 2$, or 66. Find the last number in Row 12. It is $(12 \times 13) \div 2$, or 78. Since 70 is greater than 66, Seat 70 is in Row 12.

Inductive Reasoning

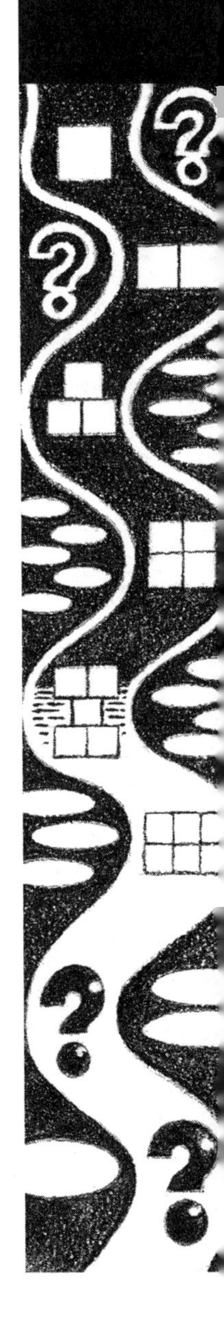

© Creative Publications

Where's the Seat? (E)

Rows	Seats
1	1 2
2	3 4 5
3	6 7 8 9
4	10 11 12 13 14
5	15 16 17 18 19 20
6	21 22 23 24 25 26 27

The diagram shows some of the seats in one section of a concert hall. The pattern of seats continues through Row 25.

1 How many seats are in Row 10? ______
How do you know?

2 Write a function rule to relate the number of seats in a row to the row number. Let R represent the row number and S represent the number of seats in a row. $S =$ ______

3 Write a function rule to relate the number of the first seat in a row to the row number. Let R represent the row number and F represent the number of the first seat in a row.
$F =$ ______

4 Marika is sitting in Seat 75. In which row is her seat? ______

Name

Permission is given by the publisher to the purchasing teacher or parent to reproduce this page for classroom or home use only.

Where's the Seat? (F)

Row	Aisle: Left	Aisle: Right
1	1	2
2	3 4	5 6
3	7 8 9	10 11 12
4	13 14 15 16	17 18 19 20
5	21 22 23 24 25	26 27 28 29 30

This diagram shows some of the seats in another section of the concert hall. The pattern of seats continues through Row 25.

1 What is the number of the seat closest to the aisle in the left section of Row 10? ______

2 Write a function rule to relate the number of the seat closest to the aisle in the left section to the row number. Let A represent the aisle seat, and R represent the row number.
$A =$ ______

3. Let L represent the last seat in the right section of any row. Let R represent the row number. Write a function rule to relate the number of the last seat in a row to the row number. $L =$ ______

4 Julianna is sitting in Seat 75. In which row and section (left or right of the aisle) is her seat? ______

Name

Permission is given by the publisher to the purchasing teacher or parent to reproduce this page for classroom or home use only.

© Creative Publications

Where's the Seat? (E)

Solutions

1 There are 11 seats in Row 10. There are 2 seats in Row 1, 3 seats in Row 2, 4 seats in Row 3, and so on. The number of seats in a row is one more than the Row Number.

2 $S = R + 1$

3 $F = [R(R + 1)] \div 2$; The first number in each row is the same as the last number in each row in Where's the Seat? (D). The rule for the first number in a row is the same rule.

4 Row 11 - Use $F = [R(R + 1)] \div 2$. Guess and check to find the first numbers in two consecutive rows so that 75 is between the two row numbers used. Find the first number in Row 11. It is $(11 \times 12) \div 2$, or 66. Find the first number in Row 12. It is $(12 \times 13) \div 2$, or 78. Since 75 is less than 78, Seat 75 is in Row 11.

Inductive Reasoning

Where's the Seat? (F)

Solutions

1 Seat 100 is closest to the aisle in the left section of Row 10. The aisle seat in the left section for Row 1 is Seat 1; for Row 2, Seat 4; for Row 3, Seat 9; and so on. The seat number is the square of the row number.

2 $A = R^2$

3 $L = R(R + 1)$; Factoring these numbers shows that the far-right seat in Row 1 is Seat 2 (1×2); in Row 2, Seat 6 (2×3); in Row 3, Seat 12 (3×4); and so on. The number of the seat is the product of the row number and one more than the row number.

4 Julianna is sitting in Seat 3 in the left section of Row 9. The number 75 is more than 64 (8^2) and less than 81 (9^2). Thus Seat 75 is either in the right section of Row 8, or in the left section of Row 9. The last seat number in the right section of Row 8 is 8×9, or 72. The left section of Row 9 starts with Seat 73. Seat 75 is the third seat in that row and section.

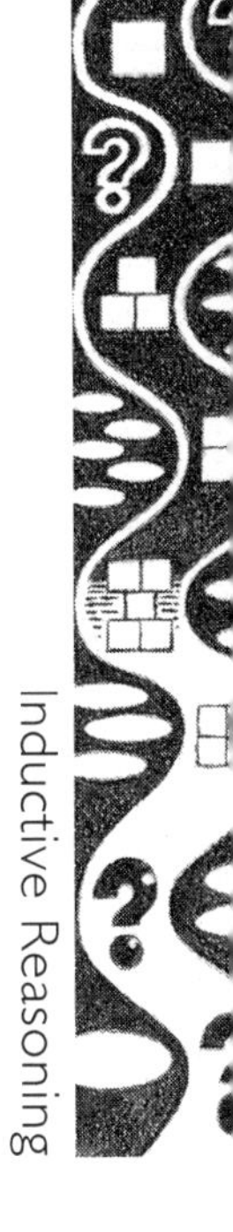

Inductive Reasoning

© Creative Publications

Certificate of Excellence

in Algebra

This is to certify that

____

has satisfactorily completed all the problems for the big idea

____

and is considered to be an expert in Algebra.

Date ________ **School** ______________________

Grade ________ **Teacher** ______________________

© Creative Publications. Permission is given by the publisher to the purchasing teacher or parent to reproduce this page for classroom or home use only.